FAITH, FAMILY & FLAG
Memoirs of an Unlikely American Samurai Crusader

Major General James Mukoyama
U.S. Army Retired

JOCKO
PUBLISHING

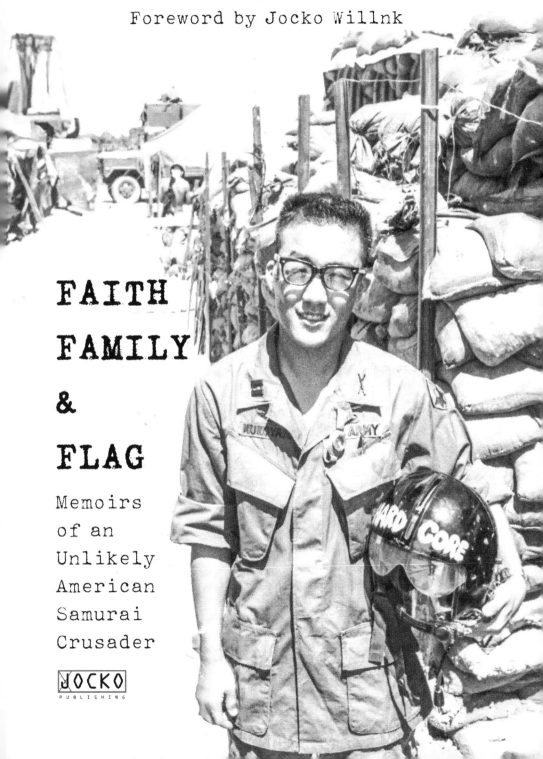

Major General James Mukoyama
U.S. Army Retired

Foreword by Jocko Willnk

FAITH
FAMILY
&
FLAG

Memoirs
of an
Unlikely
American
Samurai
Crusader

JOCKO
PUBLISHING

Faith, Family & Flag:
Memoirs of an Unlikely American Sumarai Crusader
Written by Major General James Mukoyama
Edited by Mika Burgess
Published under Jocko Publishing

US Edition Manufactured in the United States of America

For educational, entertainment and retail purposes, this
book may be ordered in bulk through Jocko Publishing.
www.jockopublishing.com

Book design and layout by Jon Bozak

Library of Congress Control Number: 2023947165

ISBN: 979-8987145234

First Case Bound Edition

10 9 8 7 6 5 4 3 2 1

1. Personal Memoirs
2. Military History

This book is dedicated to those who have enabled me to become who I am—past, present, and future. Specifically, my God for his grace for my sins; my parents for inculcating foundational values of faith and patriotism; those who served with me in our nation's armed forces for their dedication and sacrifice; and my beautiful wife, KJ for her love, understanding, and encouragement and our daughter, Sumi, son, Jae, and son-in-law, John, for loving and honoring KJ and me.

FOREWARD
JOCKO WILLINK

"James "the Mook" Mukoyama was a good man. I'd met him months before at Lewis' Training Center HQ when he was Secretary to the General Staff, and for me, it had been almost love at first sight: instinctively he'd reminded me of Jimmie Mayamura from the Raider days so long ago. As time passed, that instinct proved correct. Mook was a first-generation American of Japanese descent, whose sense of duty, honor, and country without benefit of a West Point ring was that of someone who had not yet taken it all for granted. On top of that, like Jimmie Mayamura, the Mook was a stud."

Years ago, I had read these words, written by Colonel David Hackworth, in his book *About Face*. This book has had the greatest impact on my life from a leadership perspective. Colonel Hackworth, known as Hack, was a highly decorated Army officer, who served in Korea and Vietnam, led platoon, company, and battalions in combat, and was highly regarded by his troops and superiors. But Colonel Hackworth's loyalty was to the Soldiers—the men who fought, bled, and died for this Great Nation. This loyalty eventually cost him his career—Colonel Hackworth was drummed out of the Army after he spoke his mind to the press about the poor leadership and mismanagement of the war in Vietnam.

But, Colonel Hackworth's principles and leadership values lived on in many of the men who served under him. One of those men is Major General James "Mook" Mukoyama. I was originally connected to General Mukoyama through his son, Jae, who had heard me discussing About Face on my podcast. He reached out and informed me that his father had been a company commander in Vietnam in Colonel Hackworth's 4/39th Battalion, the "HARDCORE." A few months later, I had the honor of interviewing General Mukoyama on my podcast, episode 124. This is when I learned of the amazing history, career, faith, and philosophies of General Mukoyama. But that interview just scratched the surface. This book captures General Mukoyama's entire heroic life—one of service and sacrifice, of faith and family, and of honor and humility. His example is inspiring to anyone, and his daily mantra is clear and grateful: "Every day is a great day. I have my faith, my family, and live in the finest country in the world."

But, his path to this optimism and gratitude was not an easy path. As the son of a Japanese immigrant living in Chicago, his family bore the brunt of prejudice. Some of his relatives were forced into internment camps during World War II, and his father was ineligible for citizenship until 1952. He could have found reasons to reject America and everything it stands for. But General Mukoyama was not raised to scorn America but to cherish it and the freedom it offered. He was taught that through hard work and discipline, there were greater opportunities in America than anywhere else in the world. His family raised him as a Christian, and they went to church every Sunday, always participating in and supporting their church community. Simultaneously, he was captivated by the lore of Japanese Warriors, the Samurai, and the Bushido they lived by. His devoted faith in God, America, and his ancestors' warrior spirit led him to the Army and eventually to become the first American General of Japanese Ancestry to command a U.S. Army Division.

Along this path, General Mukoyama learned to lead—in his family, in his faith, in his service to our country. From active duty in Korea and Vietnam to the Army Reserves, he codified his own leadership principles and implemented them at every level of command. This book shares those principles and gives us examples to reflect on. The book also teaches us about building relationships, the importance of hard work, and how true leadership is rooted in humility. We also learn about the service and sacrifice of the brave Americans of Japanese Ancestry who served in the valiant 100th Battalion and 442nd Regimental Combat Team. These intrepid warriors volunteered to serve America, even as some of their families were locked up in internment camps.

Additionally, General Mukoyama teaches us about subjects such as Post Traumatic Stress Disorder, Traumatic Brain Injury, and a less discussed problem, Moral Injury. The book includes backgrounds of these issues and explains what can be done to help veterans who suffer from these ailments.

But General Mukoyama does not only write or talk about problems. He has the physical and moral courage to make a stand—even at great personal risk. In fact, at the apex of his Army Career, as the youngest general in the Army, General Mukoyama did not back down when he believed that our national security and soldiers were at risk. Like Colonel Hackworth, who sacrificed his career to stand up for what he believed was right, General Mukoyama spoke truth in the face of incredible political and bureaucratic pressure and was persecuted for his actions.

It is this behavior that makes General Mukoyama a hero and an example. He managed to serve his family, his country, and his faith—all with honor by following his three leadership elements: Example, Caring, and Balance. He always set the example—to his soldiers, his children, his community. He cared about them too, more than himself. His soldiers, his wife and children, his neighbors and his congregation always came before him. And he managed this by being balanced. He logically balanced his emotions, time, and attention, allowing him to impact countless lives.

This book itself is a profound continuation of General Mukoyama's leadership. By writing *Faith, Family & Flag* and sharing his experiences and lessons learned, General Mukoyama is helping all of us live and lead better.

Thank you, General Mukoyama, for setting such an admirable example for all of us.

HARDCORE RECONDO!

Jocko Willink
September 8th, 2023

CONTENTS

EXPERIENCE

WISDOM

INTRODUCTION

One February morning in 2012, I was shaving and getting ready to go to the office when I felt a pain in my chest. It wasn't a harsh, pounding pain. I was not dizzy, sweating, or feeling feverish. I thought it was from something I had eaten the previous night and that it would go away in five minutes. It didn't.

I called my wife, KJ (Kyung Ja), and told her that this was not good, so I finished shaving and dressing. She drove me to the emergency room of the nearby hospital. I walked in and said, "Hey, I have a slight pain in my chest." They took an EKG and remarked, "You have had a heart attack." Suddenly, doctors and nurses swooped into the emergency room, and next thing I knew, I was undressed and covered with a hospital gown, put on a gurney and I was rushed into the operating room.

I have a standard daily mantra: "Every day is a great day! I have my faith, my family, and live in the finest country in the world." I say this numerous times a day, every chance I get, to friends and strangers alike.

As I was being wheeled in the corridor of the hospital on my way to the operating room, I thought to myself, "Self, can you say today is a great day?" My response was, "Absolutely, I survived two combat tours in the Army, have been blessed with a wonderful wife and family, and, most importantly, have my personal relationship with God."

In the next 24 hours, I had three operations. The first was for my heart attack. They did an angiogram and discovered my LAD, the so-called "widow-maker" artery, was 90% blocked, so they performed an angioplasty and implanted a stent. When I woke up in the recovery room, I figured that it was fixed. It wasn't.

My heart cavity started to fill up with liquid. It's called effusion. I had to go back in so a tube could be inserted in my chest to drain the liquids. It filled something like a water bottle and took 24 hours. But I wasn't finished.

My kidneys then failed. I had to return to the operating room to have another tube inserted in order to connect a dialysis machine.

Before every operation, the standard procedure at the beginning was to be asked, "What is your name, what is your birthday, and how do you feel?"

My response every time was, "Jim Mukoyama, August 3, 1944. Every day is a great day! I have my faith, my family, and live in the finest country in the world."

I can't tell you the effect my mantra had on the doctors and nurses. Before the first operation, the doctor asked, "What is your faith?" My response was, "Well, doc, since you asked, I am a Christian. Christ is my savior. You are a skilled physician with skilled nurses, but God is in control. Whatever happens, I am okay with it, so let's get on with it."

In the second operation, a nurse asked, "Where do you go to church?" My response was "Willow Creek Community Church in South Barrington, Illinois."

She immediately responded, "I do too!" No coincidence.

Finally, in the third operation, there was a male nurse in the operating room who was wearing a camouflaged gown. I looked at him and said, "You must have been in the Army."

He responded, "Yes, I was a medic."

"I just want you to know that I retired as a Two-Star General, so take good care of me…" I told him.

I faintly recall his response as I was going under. "Yes, Sir!"

How could I have responded the way I did in these circumstances? As you walk with me through my life's story, I hope to answer that question. From being an Infantryman in combat and the Army Reserves to striving to prosper as a husband, father, and disciple of God, I am grateful to have fulfilled my mission of leading and serving. My life's purpose has given me plentiful opportunities to give back to others, whether it's through my church, Army career, or the culmination of them all and my life's calling, the non-profit Military Outreach USA. I will first lay the foundation upon which I have drawn inspiration, then review key experiences — both peaks and valleys — and end with the present and future. By doing so, I aim to provide you with hope for the same peace of mind and joy.

FOUNDATION

CHAPTER 1
A NORMAN ROCKWELL LIFE: LOTTO WINNER

Growing up in the late 1940s and throughout the 1950s in the Northwest side of the City of Chicago was like winning the American lottery. My nuclear, three-generation family lived in a tenement apartment in the heart of an immigrant community, while my parents owned and operated a mom and pop gift shop on Milwaukee Avenue. The Logan Square neighborhood was truly an American environment, as it was primarily immigrants — Polish, German, Italian, Jewish and one Asian family — mine, the Mukoyamas.

At 18 years old in 1918, my father crossed the Pacific Ocean and immigrated from Japan to the United States all by himself. After my grandfather lost the family fortune investing in the sake market, he had migrated to the United States in 1901 to try and make half of it back; but he left his wife behind with five children to raise until his return. My father, Hidefumi Mukoyama, the second-eldest male, was finally sent by his mother to find his dad and return him to Japan. My father accomplished his mission, but his affinity for America kept him here for another generation. My father's outgoing and entrepreneurial spirit fit right in with America. Most Issei — the first generation of Japanese to settle in America — were introverts from rural areas, but my father, though also from a rural area, was an extrovert. He was captivated by the freedom and opportunity he saw in America. In Japan, English language study was mandatory in his elementary and high school, as well as the history of the United States. In fact, in eighth grade, my father was required to memorize Lincoln's Gettysburg address. His love for liberty was born. Having inherited his own father's traveling genes, he traveled extensively in Arizona, California, and spent numerous years in Mexico, until finally settling in Chicago in the early 1930s. For years, we thought my father's stories of exploration were all exaggerated fables. But one of my cousins had married a woman of Mexican heritage, and she recognized the lyrics and tune of a Spanish ballad my father had been singing for years. Since then, his stories have become legendary.

After finally settling in Chicago, my father met my mother, Miye Maruyama. My mother was living with her parents in Oklahoma and considerably younger than my father by over a decade. Although they were introduced and connected by another Issei, their relationship was not an automatic arranged deal. My mother was Nisei, which meant she was part of the second generation to live in America, her parents having immigrated from Japan. Being born and raised in the States, she had to be convinced to marry him, which luckily my father was able to accomplish. They were married in 1936 and set up an apartment in Chicago. A year later my brother, John, was born. I followed seven years later, on August 3, 1944.

My parents made their home in a tenement apartment on the Northside of Chicago, in a close-knit community called Logan Square. We lived on the top floor, above an Italian family and English and Eastern European residents. Everyone knew and liked each other, and my parents became active in the community and in our local church. We were fully assimilated and accepted in our diverse immigrant community. We knew everyone in the neighborhood by name, and they knew us. Doors stayed unlocked, and mothers acted as the neighborhood police. In those days, most of the mothers were homemakers raising their children, but since we had a family business, my mother worked six days a week at the store.

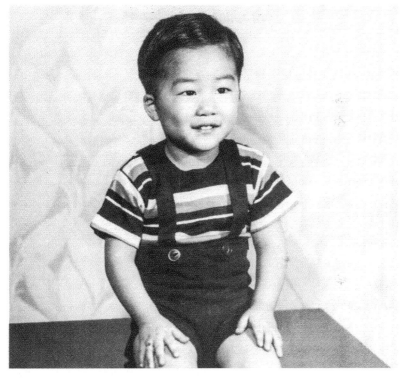

A bright-eyed, two year old in Chicago, IL. 1946

After the War, my Grandparents came to live with us, and my grandmother cared for my brother and I during the day. This strong nuclear family and tight-knit community experience left a strong impression on my life. Since our parents were busy holding onto responsibilities at the store, my brother and I were responsible for cleaning the house. We never complained about these chores; rather we'd turn the tasks into a fun game. We split up the chores and rooms and raced to see which one of us would finish dusting, sweeping, and cleaning first. My brother and I had a large age gap, so our split tasks were one of the few activities that brought us together. I looked to him as an example, although he often looked at me as his kid brother. Our shared responsibilities bonded us and instilled a sense of pride and duty to our family too.

Every Sunday, we would put on our Sunday best and walk together to church as a family. Our church, Avondale Methodist, was three blocks from our home, and my early years were centered around the church. My older brother and I were baptized and confirmed in that church, and we bonded with our God and community inside those sacred walls. I joined the junior choir and sang a solo on the local WGN-TV channel on a program called *Faith of Our Fathers.* Our Cub Scout Pack and Boy Scout Troop were sponsored and hosted in our Church too.

My mother and Grandparents were strong Christians, and they imparted the value of faith on the rest of us. Although my father was not raised as a Christian in Japan, he followed and became an active member of our church. His father, my paternal grandfather, was also a Christian, which was rather unusual, since Christians in Japan are few and far between. My family created a foundation for us to accept God into our lives, and faith preserved the strongest bonds between us.

Throughout the week, our family operated the gift shop on Milwaukee Avenue in town. Mom and pop shops have been cited as the backbone of the American economy, and I was a first-hand witness to this truth. Our parents worked six days a week, from 9 to 5 and 9 to 9 on Mondays and Thursdays, to provide for our family and stimulate the local economy. Through my parents' entrepreneurial and diligent spirit, I observed the hard-working determination that connects most Americans. The Logan Square shopping district epitomized an American Main Street, but in the diverse urban atmosphere of Chicago. Our small retail shop was stocked with figurines, lamps, vases, dishes, and other knickknacks that were imported from Europe and Asia. As I watched my parents strive to sustain this gift shop, they were teaching me a lesson in American values. Although the store never reaped substantial profits, our family business gave me an invaluable lesson in the American principles of hard work, individualism, and enterprise.

My parents' example instilled a life-long work ethic in my brother and me. When I was in elementary school I had a paper route, and I started working in the warehouse district as soon as I could get a part-time summer job. A.C. McClurg's was a large national wholesaler, and the warehouse occupied an entire city block near the Navy Pier in Chicago, built right after the Great Chicago Fire of 1871. I worked in the book department, which had thousands of titles in stacks occupying an entire floor of the warehouse. McClurg's had the contract for all the public libraries in the State of Wisconsin. The first summer, I was given the job of a "picker" and had to fill an invoice by pulling heavy books, packing them for shipment, and carrying them to be sent out on a conveyor belt. I suffered through that summer, as the tough manual labor in the old, dusty, non-air-conditioned warehouse triggered my seasonal hay fever. I was so grateful to have a job and earn $1.15 an hour, but I figured there must be a better way to earn a salary. I noticed that typists, or "billers," sat comfortably in chairs at the end of the conveyor belt and took invoices, typed in quantities and prices which were automatically calculated and printed with the total price per item with applied discounts and a final total amount. This no-brainer alternative would earn me my paycheck, while requiring less elbow grease and sweat.

When I returned to school, I took the initiative and immediately signed up for a typing course. I became quite proficient and could type about 65 words per minute, and the class even came with two ancillary benefits. I was able to type all my school papers, which eventually came in handy in college and graduate school. But more importantly at the time, I was the only guy in the typing class, so I was able to meet girls.

As a senior in high school, I worked hard from 5 pm until 10 pm, Monday through Friday, and then 9 am to 5 pm on Saturdays, in order to earn money for college. My parents could not afford to send us to college, so we had to save for our tuition. But my brother and I never thought to complain. It was our lot in life, and many were worse off than we were. We had much to be grateful for, so we kept our heads down and hammered away. My senior year ended up being my best academically, as I achieved straight A grades in all my subjects. My demanding schedule forced me to learn time management, balancing my job, school, and social activities. My hard work paid off.

Neither of my parents went to college but graduated with highest honors from the school of hard knocks — life. They were both high school graduates, which was basically equivalent to a college degree today. Yet, they were entrepreneurial, and our small retail family business provided a steady income for our family for over three decades, until the big box stores came in and drove the mom and pop shops out of business.

We were rich in faith and family. Although we never owned a home during this time, I never felt poor. Our cups truly ran over. Our parents stressed to my brother and me to always be proud of our Japanese heritage and never shame the Mukoyama name or the Japanese race. But they also always emphasized that we should be proud of being citizens of the United States of America. Coming from immigrant backgrounds, they were forever grateful for the freedoms this country offered. We had opportunities afforded by our nation that were not available in such abundance anywhere else in the world.

The day my father became a U.S. citizen was one of the proudest days of his life. During the proceedings, the immigration judge incredulously asked, "Why did it take you so long to apply?" Offering a respectful impromptu civics lesson, my father replied that that year, 1952, was the first year that Japanese immigrants to the United States were authorized to become naturalized citizens of this country, due to the passing of the McCarran-Walter Act. To my knowledge, he was never bitter for having to wait so long. He worked hard to be a strong contributor to our community. He belonged to the Chamber of Commerce, was active in our church, and was a strong supporter of our Boy Scouts program. He also supported members of the Japanese American community in need of hope and help, prior to, during, and after World War II. He realized how blessed he was to become a citizen of this nation, so he worked to be the best citizen he could be. I never took my birthright citizenship for granted, and it's the reason that I say, "When I was born here, I hit the lotto."

When I've made speeches in my older years, I often reflect and mention that I have led a Norman Rockwell life. Norman Rockwell was one of our most beloved American artists of the twentieth century, and from 1916 to 1963, he created over 300 covers for the popular Saturday Evening Post magazine. He was famous for his detailed depictions of everyday life in America, emphasizing the blue-collar inner city and rural America landscape. From high school dances to breakfast table conversations, he illustrated the modest moments that often exist only within the confines of our memories. Like millions of Americans, I was able to relate to his covers and depictions. His most well-known paintings were the so-called *Four Freedoms* which he drew in 1941 that reflected the themes of President Roosevelt's State of the Union address: *Freedom of Speech, Freedom of Worship, Freedom from Want,* and *Freedom from Fear.* Whenever I recall these paintings, especially *Freedom from Want,* also referred to as the "Norman Rockwell Thanksgiving" painting, I say to myself, "This was our family."

On Christmas and Thanksgiving, my family would gather together and feast. For the other 363 days of the year, my parents would always be busy operating the shop, but holidays were special. Christmas and Thanksgiving placed God and Country at the center of our family. When I look at Norman Rockwell's *Freedom from Want,* I can see my mother serving a delicious turkey and mashed potatoes, and my family gathered around the table together to give grace. On these holidays, I learned the importance of faith and family. More so, I became aware of how blessed I was to live in the United States, a country where we dedicated special occasions to count and celebrate these blessings. The seeds of my daily mantra were planted on these holidays of my childhood, and this gratitude blossomed as I celebrated with my own wife and children years later. Today, a copy of *Freedom from Want* is hanging in a prominent place in our home.

We were a close family, bonded by shared religious devotion and mutual responsibility, staying optimistic and grateful for one another and our blessings. We were a critical component of the quintessential urban American community. Thus, the foundational values for my life were set and began to shape me. Throughout my life, these ideals have been re-emphasized in my life experiences, little moments that I refer to as "God things".

CHAPTER 2
LIFE PURPOSE: SERVING OTHERS

When I was 9 years old, I asked myself, "Why am I here?" As an adult, I have questioned my motives behind what caused this important internal conversation at such a young age. For the life of me, I can't recall the causes, but I remember my conclusions clearly.

I am here to make the world a better place by helping others. I do not need to get famous for my benevolence or do something grand like finding a cure for cancer, but I want to make a positive impact. If I could say that the world was a better place because I was here, I would have fulfilled my reason for having lived. I reasoned that if everyone did the same, the world would continuously improve for all.

The best way to accomplish my life's purpose is by serving others. This concept was first instilled in me by the Boy Scouts of America's "Scout Law," which includes the vow "to do my duty to God and Country" and "to help other people at all times." While attending Carl Schurz High School in Chicago, as a junior, I was awarded the Outstanding Service Award because of my activities as a volunteer in the school library, secretary of the band, chaplain of the Key Club, member of the Tri-Hi-Y club, officer in the ROTC, and member of the Laurels Academic and National Honor

Societies. Rather than playing sports, I believed my time was best invested in helping people and organizations. Later in life, my dedication to service was further confirmed by the Preamble to the Constitution of the American Legion, the nation's largest Veterans Service Organization, which states, in part, "for God and country, we associate ourselves together ... to inculcate a sense of individual obligation to the community, state and nation." And who was the best servant of all time? The one who voluntarily gave up his life to save mankind: my Savior, Jesus Christ. He set the standard by saying that He did not come here to be served, but rather to serve. This attitude is the one I mirrored in the military and my civilian life.

By serving the people under us, the traditional leadership pyramid is turned upside down. Instead of the leader at the pinnacle being supported by the followers on the bottom, the followers are instead at the top being served by their leaders. The night He spent His final meal with His disciples, Jesus put on a serving towel and washed their feet. Servant leadership is an absolute devotion to your people, and Jesus demonstrated and accepted this responsibility when He died for all of our sins.

Due to my modest achievements in the realms of service, I've been invited to give numerous presentations on leadership. Now, I am quite simple-minded, and I've boiled down the most important concepts to their basic elements. Accordingly, when I address the topic of leadership, I mention three key principles. My leadership principles have been distilled down to three key elements: Example, Caring, and Balance.

In any organization, leaders must lead by personal Example. It's not "do as I say" but "do as I do." Subordinates can smell a fake leader from a mile away. This lesson was demonstrated by the man who set the example for me and so many other outstanding servicemen. I had the honor to serve in Vietnam under one of the finest combat leaders in the history of the United States Army, LTC David Hackworth. I was blessed to have learned his leadership principles from him directly.

LTC Hackworth set an example every time he stepped on the battlefield, meriting our respect and showing us how much he cared. I was a company commander in a battalion he led in the Mekong Delta in 1969. LTC Hackworth was a Korean War Veteran who was awarded his eighth — no, that's not a misprint — Purple Heart Medal for wounds sustained in combat while he was our commander. While he was monitoring a firefight in his Command & Control helicopter, he noticed some of our soldiers were in trouble and seriously wounded. He ordered his pilot to land the chopper in the middle of the firefight to extract and save the wounded soldiers. The reluctant pilot needed some convincing but eventually landed. The small C&C helicopter was able to load the two wounded men but there was no room for Hackworth. He ordered the pilot to lift off with him standing on the skids. He got hit in the leg, earning him his eighth Purple Heart. After that incident, Hackworth could have ordered us to walk through a wall of fire, and we would have simply asked, "Where do you want us to go?" By setting an example of a selfless and fearless warrior and leader, LTC Hackworth earned our lasting trust and admiration that went beyond the battlefield and into our hearts. Leaders must set the example physically, mentally, and ethically.

Next comes Care. Leaders should care for their team through tuning into the truth, empathizing with their positions, and fostering relationships. Everyone starts their careers at the bottom of any organization, and leaders must always remember where they came from. They must also recall the servant leadership principle mentioned earlier. In order to serve them, you have to know your subordinates and their life situations. They have to perceive that you truly care about them as individuals, rather than treating them like pawns to be maneuvered for your personal career advancement. If they recognize your sincere concern for them, they will reciprocate in kind because of the trust that will be generated in that relationship. When you are a leader, you must take the time to roll up your sleeves and spend time getting your hands dirty, whether it be on the ground with soldiers or on the production line in a factory.

As you get higher and higher in an organization, the bureaucracy will create barriers to halt you from maintaining contact with those on the ground. Often it is well-intended to save you valuable time, but don't fall into that trap. Not only does it isolate you, but disengaging with your team makes you fully dependent on the information from your staff. In the military, I employed a unique technique to avoid both. Prior to visiting a unit in my command, my staff would give me a detailed briefing about the unit that included their personnel, equipment, training, and readiness status. We had a ton of data in all these areas, and I would be presented with how that unit compared to all the other units in the command.

So, what did I do when I made the actual visit? I would have my senior enlisted soldier, normally a Sergeant Major, get the personnel roster and randomly select five or six individuals for me to speak to. No senior officer or senior non-commissioned officer would be selected because they know what the commander wants to hear. I required that I heard from at least one woman, at least one minority, and at least one non-mainline occupational specialty soldier. I wanted to hear perspectives from a wide range of people, so that I accounted and cared for everyone under my command. I would then interview each individually, with the Sergeant Major as a witness.

I would start by telling them to relax, which is not easy for a lower-ranking soldier to do in the presence of a General Officer. In the military, the unit commander, normally a Captain, is considered a mini-God and referred to as "the old man." One rarely, if ever, gets to see a General, much less speak to one. I understood this and emphasized that our conversation would be completely confidential and undisclosed. Among my various questions, I wanted to understand how leadership affected the unit and how it could be improved, both on an overarching individual level. I posed questions like: Have you been counseled by anyone regarding your performance and professional education opportunities? On a scale of one to ten, how would you rate the morale in this unit, and how would you rate the leadership in the unit? If you were king or queen for a day, what is one change you would make? Even though the discussion only lasted 10 minutes, I imagine it seemed like eternity to the interviewee. After just one hour of interviewing these individuals, I had a pretty accurate hand on the pulse of the unit. Their perspective was so important because there were times when it was 180 degrees different from what I had been briefed on by my staff. I discovered that the best way to care about your subordinates is by demonstrating a sincere interest in the truth.

And finally, leaders must maintain Balance in their lives. Perks and praises come along with success, and it is tempting to get caught up in all the glory. In fact, it can become an addictive focal point of your success. At that time, it becomes tempting to ignore the most important human relationships in your life: your family and significant others. I always advised my associates and team members that they would be doing our organization no good if their family life was falling apart at the seams. If your personal life is falling apart, the organization cannot thrive. Compliments and benefits can perpetuate your ego, but smugness will not do you, or your team, any good. Whether in work and family life or ego versus humility, balance in these realms has always reaped the most success in my leadership experience.

Being a servant leader has two ancillary, but profound, benefits. First, the more you do it, the happier you become, because of the personal satisfaction one receives from serving others. Second, it encourages the recipient of your service to also help others.

Having a servant attitude also engenders humility. My wife, KJ, has done a remarkable job in making sure my head did not exceed my hat size, which was not easy for a type-A guy like me. Further, my ego has often been checked when it's getting out of hand. When I was selected for promotion to General Officer, the Army had a "General Officers' Orientation Course," nicknamed the "Charm School," for new Generals and their spouses. For one week, we attended meetings and briefings at the Pentagon, State Department, and Capitol, and we were literally wined and dined every night. We were scheduled to visit the Vietnam Veterans Memorial wall one night, which I had successfully avoided visiting up to that point. When we got off our bus, there was a helicopter above shining a light and illuminating the wall against the dark night sky. We were so impressed that they would go through so much trouble for our group, but then quickly learned it was a serendipitous moment. A nearby 7-11 had been robbed, and it was a police helicopter searching for the robbers. That memory has stuck with me as a lesson to always stay humble. The visit ended up being a very cathartic experience, as I visited the names of my comrades on the wall and remembered their sacrifices.

In the process of writing this book, numerous revelations emerged regarding my life. As I've reflected on my attempts to live up to my life's purpose, I've found that my experiences are all interwoven by a thread of servant leadership interlaced with a drive to set a positive example, provide care, and create balance. My Christian faith is the cornerstone of my dedication to service.

My nine-year-old self asked a difficult question. The answer to that question has been clearly reflected and revealed throughout my life.

CHAPTER 3
UNDERSTANDING MY ROOTS: EAST OR WEST?

My story begins at the end of World War II, when my two foundational origins were at war. The animosity between Japan and the United States left me at a crossroads. But I followed the trail my father and mother paved to reconcile my roots and forge my own path.

I had an Issei (first generation immigrant) father who inculcated the value of Japanese culture into my brother and me. My parents demonstrated the core cultural tenets through how they carried themselves and lived their lives. They instilled the concepts of *giri* (honor), *zhugi* (loyalty), *kenshin* (dedication), *on* (debt, lifelong), *keiro* (respect for elders), and *gaman* (perseverance).

Although my parents never taught us these Japanese words outright, their characters gave us a valuable model to learn from. For instance, they demonstrated *giri*, *on*, and *gaman* during a time of extreme crisis in our family. In the 1960s, the growth of the large superstores like K-Mart crushed the competition of small retail shops, and our family business faced several consecutive losing years. We could no longer compete. It would have been easy for our family to declare bankruptcy as many did, but my father refused. He told us that the creditors had supported us in good faith, and he would not leave them without repayment. So in his mid-60s, he went back to work as a clerk for an importer of gifts. My father had helped this merchant start his business decades earlier, and the owner never forgot. Although it took numerous years, all of our creditors were paid back. He persevered with honor, to honor his debts. The Mukoyama family name was unstained.

These Japanese cultural concepts were part of the Samurai tradition of *Bushido*, or the "Way of the Warrior." *Bushido* is the moral code that guides samurais' way of life. In Japanese culture, the warrior was at the top of the social ladder, unlike in China where the scholar held that position in society. Although Japan has not had an offensive army since World War II, the warrior ways are still deeply entrenched in how Japanese people carry themselves and treat others. The stories about these warriors have become national legends.

The admiration of *Bushido* is highlighted in a famous historical event that has become Japanese folklore. A powerful samurai clan, led by Tokugawa Ieyasu, ruled Japan for over 250 years, from the early 1600s to the mid-1800s. During this time, the samurai reigned supreme throughout the country. The story is called *Chushingura*, or "The Tale of the Forty-Seven Ronin (masterless samurai)." In short, it is a classic story of an evil Japanese daimyo, or lord, who insulted another daimyo to the point where the latter drew his sword in the shogun's palace and cut the offending daimyo. In order to protect the shogun, it was a capital crime to draw your sword in the palace, and therefore, the offended daimyo who drew his sword was required to commit *hara-kiri*, a ritual suicide by disembowelment. His retainer samurais were disbanded, and his land was given to the evil lord. But *Bushido* code also required the samurai followers to seek revenge on behalf of their dishonored lord. The evil lord knew this, so he employed twice the number of samurai to repel any attack. The retainers of the "good" daimyo concocted a secret plan to split up and to throw the evil lord off their trail. Many left their families and became *ronin*, masterless samurai, acting like drunkards and exhibiting other bad social behaviors. They became social outcasts and endured ostracism and insults. But they made a sacred pact signed in blood to regroup at a designated date about a year later to gain revenge. On the appointed night, they regrouped and attacked the evil lord's castle. Although heavily outnumbered, they overwhelmed all opposition and caught and respectfully offered the evil lord the honor to commit *hara-kiri*, but he was a coward. The masterless samurai executed him. They cut off his head and carried it through the streets to the cemetery where their lord was buried, so his spirit could be at rest.

Overnight, the samurais became national heroes. But there was one problem. By executing a lord higher than they were in the samurai social structure, they were also required to commit *hara-kiri*. The public clamored for clemency, but the shogun refused, so all 47 went to the cemetery and committed *hara-kiri*. They are revered to this day in Japan, and the Japanese movie industry has produced movies about them for decades.

Japanese films gave me an early glimpse into my ancestry. My family's special outing was going to the only place in Chicago that showed Japanese movies, which was a private school in the city called Francis Parker where they would show a double feature of Japanese films. The first movie would tell a traditional samurai story, like *The Tale of the 47 Ronin*, whereas the second would be a modern storyline that highlighted Japanese values. For example, I remember a film about an older gentleman who saved a community park from being torn down and turned into a parking lot. All these films spotlighted an important lesson that tied back into a core cultural concept. Outside of my family, these movies gave me the first understanding of my Japanese heritage. I also tried to explore Japanese traditions, language, and culture on my own. When I was 11 years old, I rode my bike to the only Buddhist temple nearby, hoping to learn the Japanese language with assistance from a Buddhist priest. Especially after World War II, very few people wanted to learn Japanese. But I took the initiative, biking 30 minutes there and back. However, the Buddhist priest never showed up. My quest had to be postponed, but I was still determined to learn more about my Japanese heritage. Eventually I would gain a full Far East education at my university, but in the meantime, I came to understand the intersection of my Eastern and Western Roots as a member of the Japanese American community.

To fully understand the unique experience of Japanese Americans in the United States, a brief immigration and language lesson is in order. The pattern of Japanese immigration to America is an anthropologist's dream because the circumstances resulted in three distinct large generational cohorts. The word for generation in Japanese is *sei*. The numbers one, two, and three in Japanese are *ichi, ni*, and *san*. Japanese immigration began in the late 1880s, first to Hawaii, and then to the mainland of the U.S. But this migration was shut down in the early 1920s due to a combination of declining labor needs of America and subsequent anti-Japanese legislation. Accordingly, a clearly defined cohort of young adults of the first generation — or *issei* (combination of ichi and sei) — entered into the United States. Due to laws prohibiting interracial marriages, as well as Japanese cultural norms, the issei married other issei, and they all had babies around the same time, resulting in *nisei* (combination of ni and sei). This pattern was repeated with their children, resulting in the sansei. As a result, in the 1960s, when one encountered a Japanese American, you would generally know if they were issei, nisei or sansei by their age. Grandparents were issei, parents were nisei, and kids in my generation were sansei. But there were a few exceptions when issei would marry nisei. In Japanese, the word for half is *han*, and therefore, their offspring were referred to as *nisei-han*. My brother and I were proud nisei-han. We also wore the Mukoyama name with pride. Our family name reflects the rural area of Japan where my father was born. In Japanese, *muko* means "over there", and *yama* means "mountain." We lived in a rural area in Yamanashi Prefecture, outside of a city called Kofu. My father could see the beautiful Mt. Fuji from his backyard, hence the derivation of our family name. When people ask me what my name means, I respond with a poetic, "yonder mountain." My father's name was "Hidefumi"; it is also my middle name. *Hide* means "English" and *fumi* is "literature." Coincidentally, I majored in English Literature for my bachelor's degree at the University of Illinois. Our Japanese last name has kept us connected to our family history, even as we followed new American pursuits.

We were constantly reminded of the strong cultural foundations of our ethnic heritage, of the 26 unbroken lines of the Imperial family in Japan. My father's family in rural Japan could trace our family tree back hundreds of years. The local village cemetery records were maintained by the temple monks, and they preserved the long history of the Mukoyamas. Virtually all of our relatives were buried in that cemetery.

My father made us feel so special because of our 3,000-year history. I secretly felt sorry for my Caucasian friends because they did not have this extensive and eminent family background. I felt truly blessed to have such a rich cultural heritage.

21

Although my mother was a nisei and had only lived in the Midwest, she instilled the values she gained from being an American immigrant into my brother and me. My mother was born in Madison, Wisconsin and grew up in Oklahoma. Her parents immigrated from Japan. Like all of us who grew up in the U.S., she was raised with the American spirit of individualism and adventure. Having survived the Depression in the Dust Bowl of America, she had been toughened by her conditions into a true stalwart. Her parents were Christians, and her faith was a critical component of her life. As a high school senior, she won a state-wide essay contest all about saving for the future that quoted Scripture verses. She was an outstanding student academically, and one of her proudest accomplishments was becoming a member of the National Honor Society in the second half of her junior year, the earliest semester one became eligible for membership. Once I was in high school, my mother bargained with me that I could receive her Honor Society pin, but only if I qualified my junior year as well. I was successful and became a member, and I still have her NHS pin, which is one of my most prized possessions. My mother inculcated my father, my brother John and me into her Christian faith, love for America, and a tremendous work ethic.

With our family business, my mother worked six days a week at the store, then she did the laundry, cooked our meals, ironed our clothes, and checked our homework. She also had her parents living with us for several years, and she was caring for them in any of her spare time.

My grandparents' presence was always a welcomed addition, never a burden. Some of my fondest memories were of my grandmother doting over my brother and me. My grandfather, who was tall for Japanese, always appeared to me as a Lincolnesque figure. He didn't speak much English, but when he spoke, everyone listened. I never saw him lose his temper. We lived on the third floor of an apartment building, and I remember how he used to throw breadcrumbs on the roof of the house next door to feed the birds. Growing up close with my grandparents was a cherished experience, and our three-generation household followed Japanese tradition too.

For too many Americans of Japanese ancestry, our history was shrouded in injustice due to their incarceration in internment camps. During World War II, Japanese Americans were placed in concentration camps with barbed wire fences and armed guards with machine guns in towers facing in, not out. These camps were euphemistically referred to as "relocation centers", but these concentration camps seized thousands of Americans' liberties and property and thwarted their pursuit of happiness. Over 120,000 persons of Japanese ancestry, two-thirds of whom were American citizens, were evacuated from the Western states and moved to the interior without any due process, simply due to the color of their skin, by President Roosevelt's Executive Order 9066, signed in February 1942.

Our immediate family was not placed in the concentration camps because our neighbors came forth to support us. Since my parents had assimilated so well into our community, our neighborhood sent a telegram to our congressman vouching for the loyalty of my father as an American. They did not know that he was ineligible to become a citizen, due to the U.S. having discriminatory laws against people of Japanese descent. But they knew our family character by witnessing our devotion to our faith and community. Actions speak louder than words, and both my family and my neighbors' loyalty to one another shined through.

These hopeful experiences allow me to place my Americanism at the forefront of my identity. I personally prefer to describe myself and others as American of Japanese Ancestry (AJAs). Although I also refer to our community as Japanese Americans, I believe our United States citizenship should prevail. The title, American of Japanese Ancestry, empowers both my nationality and ethnic origin.

However, this is not to say that Americans of Japanese Ancestry did not experience discrimination. There were incidents of racial taunts and threats. My father often stood up for victims of discrimination, and he recalled once getting involved on behalf of a nisei nurse who died in Chicago. Because she was of Japanese descent, her family could not find a cemetery that would provide a burial place. Keep in mind she was an American citizen, not an enemy alien as those non-citizens were categorized. My father and other isseis drove all over the Chicago metropolitan area, until he finally found one cemetery, Montrose Cemetery, on the far northwest side of the city, that would accept her body. The Japanese Mutual Aid Society purchased a large section of the cemetery and built a large mausoleum to honor Japanese, regardless of nationality. For decades, virtually all AJAs were buried in the Montrose Cemetery. To this day, there is a community ceremony conducted at the cemetery on Memorial Day that attracts large crowds and involves Buddhist temples, Christian AJA churches, the Japanese Consulate, and the Chicago Nisei American Legion Post 1183. There is a Veterans Memorial near the mausoleum that honors the nisei soldiers who died in World War II. My family's contributions to serve and commemorate the lives of these Americans of Japanese Ancestry is part of our family history I hold dear.

Towards the end of the War, the Japanese Americans in the camps were relocated to urban areas in the interior of the country. Chicago became a primary reception area because of the availability of jobs. My parents, along with my father's brother, Teruo

Mukoyama, and his wife, Helen (Kiyoko), worked alongside a small number of Japanese to support those coming out of the internment camps. They formed the Resettlers Committee and provided apartments and jobs for those coming out of camp to Chicago.

Mukoyama brothers at family cemetery (L- James Mukoyama Sr., Teruo Mukoyama). Kofu, Japan. 1978

During the War, our home was an unofficial USO for Japanese American soldiers on leave. My parents were models of service to others. My father was also active in establishing the previously mentioned Japanese Mutual Aid Society for those Japanese in need of community services. Throughout this time, my parents remained patriotic to the United States of America.

After the War, my Uncle Teruo represented the AJA community in America, presenting financial aid raised by the AJAs throughout our country for the Japanese people suffering from the atomic bombs and fire bombings of Japanese cities. These two weapons were devastating to civilians, and the firebombing actually caused more casualties than both A-bombs. Teruo was honored to receive an audience with the Japanese Emperor in Tokyo and was appointed to serve on General MacArthur's occupation staff.

For their outstanding community services to the Japanese people in Japan and in the United States, both my father and uncle were honored by the Emperor with the awarding of the Order of the Rising Sun, which is commonly referred to as a *kunsho.*

However, the United States had stripped the Constitutional rights, freedoms, and property of thousands of Japanese Americans, and the survivors deserved some sort of redress and acknowledgement. A federal commission found that this injustice was due to wartime hysteria, race prejudice, and failed political leadership. My aunt Helen was a licensed social worker who testified in the Chicago Congressional hearings. Eventually, the findings of these hearings would result in the Civil Liberties Act of 1988. In this Act, the United States Government officially apologized for the injustice of the concentration camps and issued reparation payments of $20,000 to each person still living who was actually imprisoned. It should be noted that no payments were authorized for future descendants who did not personally experience the injustice. For those who lost everything — their homes, property, businesses, and reputations — the compensation was a token payment, but the apology restored

26

their dignity and honor (*giri*). To many Japanese Americans like my grandmother, this restored respect meant everything.

My maternal grandmother, Rui Maruyama, was one of the thousands who had been forced into these concentration camps. She was living in Pasadena (yes, "the little old lady from Pasadena" for all you Beach Boys fans) in her 90s when the arduous legislative campaign for the Act began. She vowed to live to receive the apology and payment. She did, and eventually passed away at well over 100 years of age.

As a result of these two major cultural influences — the Eastern from my father and the Western from my mother — my early childhood through my adolescent years embodied the American immigrant experience. I shared this experience with millions of other American families. Although we must come to terms with our governments' mistakes — such as Executive Order 9066 — this country has provided my family and me with boundless opportunities. I have been blessed to be a citizen of the greatest land of freedom and opportunity in the world, with pride in my family's past and ancestry.

As my father emphasized and exemplified, we should take pride in our ethnic heritage, but our future efforts must be dedicated to the country of our citizenship — the United States of America. My family has stood firm in this nation's core principles — faith, liberty, and an American citizenry from a diversity of nations that forms the finest country in the world. My deeply planted roots were neither Eastern or Western, but American.

CHAPTER 4
JUDEO-CHRISTIAN EXPERIENCE — GOT 'EM COMING AND GOING

Religion was the bedrock of my community. In a city defined by immigrants' different ethnicities, we searched for commonalities to bond us together. The residents of Logan Square all frequented the same local grocery stores, cleaners, taverns, and barber shops. But the various churches, synagogues, and temples truly bonded our community together. The houses of worship were all the center of activities in our neighborhoods.

As a young American growing up in the 1950s, my experience was stereotypical, especially in regard to my church, Avondale Methodist Church. I was baptized and confirmed there, and our Cub Scout pack and Boy Scout troop were sponsored by the church and met there too.

As young boys whose habitat was the concrete jungle, Cub and Boy Scouts provided an uncharted terrain to explore our love of faith and country. Membership in the pack and troop was not limited to our church members. You just had to be a young male seeking the scouting experience with other boys. This environment was especially attractive to those of us who lived in the inner city. We had only experienced the asphalt jungle without large backyards or lawns, so Boy Scouts provided an opportunity to discover the great outdoors.

Our Scout meetings were conducted in our church, and faith was an integral part of Scouting. It was ingrained in us by the Scout Oath, which stressed "to do my duty to God and Country" and to obey the Scout Law. These laws instructed us to be "reverent", defined as "Be reverent toward God; Be faithful in your religious duties; Respect the beliefs of others."

Our church was blessed with two men that became critical to the success of our scouting program. Our Scoutmaster, Woody Allen, was a military Veteran from rural Kentucky. He was an expert in hunting, fishing, and camping, and he was dedicated to educating troops about the outdoors. The second was an unknown benefactor who donated a wooded plot of land in an area north of Chicago, which today is an upscale suburb. Our troop cleared enough land to make a campsite with a fire pit in the middle and wooden platforms for individual tents surrounding the fire pit. We even built a wooden fence around the campsite out of the trees we had cut down. There was no electricity or running water, so it was us and Mother Nature. We loved it. We cooked our meals over the fire, which taught us resiliency and responsibility. We slept outside in the fresh air and under the stars in our two-man "pup" tents and sleeping bags. At night, we took turns staying up throughout the night for "fire watch" to make sure that the fire did not spread. For us city boys, this campsite was a great experience to enjoy nature, the stars in the evening, and the fresh air.

The values emphasized in the scouting program were a perfect reinforcement of the Japanese culture and Christian faith I was simultaneously receiving at home. The Scout Law emphasized that a Scout was trustworthy, loyal, helpful, friendly, courteous, kind, obedient, cheerful, thrifty, brave, clean and reverent.

When I was twelve, I had the lucky opportunity to attend the Boy Scouts National Jamboree in Valley Forge, Pennsylvania, with 52,000 of my best friends camping out in tents. I was the only Scout from our troop to attend. Most could not afford the expense, but my parents made sacrifices to send me. I was so grateful because the National Jamboree was a glimpse into what the United States was like outside of my small scope of Logan Square. Our Scout district formed a composite troop with boys from many individual troops. For the first time, I associated with Black and Hispanic American boys, and we bonded very closely while preparing for the Jamboree and living together for the week-long encampment. Although we all had different appearances and came from different areas, we found similarities and formed quick friendships.

Detail from Boy Scout Jamboree, Valley Forge, PA. 1956

While at the Jamboree, we visited the historic sites of the

founding of our nation in Philadelphia. In the evening, we had religious services with all of the Scouts together. The logo and patch for the Jamboree was President George Washington kneeling in prayer during the harsh winter of 1777 at Valley Forge. Washington's leadership improved the troops' professionalism and raised the morale of the fledgling Continental Army for the long campaign ahead. This image symbolized the lessons and characteristics that the National Jamboree was trying to train us to emanate. I met Scouts from all parts of our country, and even international Scouts from Japan. The National Jamboree, and my entire time at Boy Scouts, highlighted two of my core values, Faith and Flag — but there was one day three years earlier that was the turning point for my entire life.

In the summer of 1953, the most important event in my life occurred. My mother's sister invited me to join her, her daughter, and another cousin at a Moody Bible Day Camp in the city. It was a fun day for us kids, with games and arts and crafts. I made a cover for my Bible, which included imitation alligator and leather stitching. Towards the end of the day, there was a tent set up outside for a service, and the minister gave a passionate sermon. At the end of the service, the minister issued an invitation for anyone who had not accepted Jesus into his heart to step forward. I was sitting in the back and listening to his message, until I decided to get up.

I went to the altar and opened my heart up to the Lord. I have walked with Him ever since. Have I had failures, disappointments, betrayals, and stumbled in life? Absolutely, as you will learn, but throughout I knew that God was with me and I was never alone.

My coming to faith did not have a dramatic "road to Damascus" story like the Apostle Paul. But it was the beginning of what I refer to as "God things" in my life. These pivotal moments are when the

Good Lord has provided key experiences and individuals at crucial times that profoundly influenced my life.

Speaking of the Apostle Paul, formerly Saul, I have had the good fortune to have Jewish friends and mentors, and associations with the Jewish community throughout my life. These good-hearted individuals have been guides along my path, and it all began with my birth. Our family doctor was a Jewish man named William (Wolfgang) Shapiro, and his office was on the same block of Milwaukee Avenue as our retail gift shop. In those days, doctors made house calls, and people did not go to hospitals unless medical treatment was required. Doctor Shapiro brought both my brother and I into the world and took care of all our medical problems, including my appendicitis and the removal of a potentially serious tumor from my head.

Doc Shapiro was also a very close friend of my father, and a caring and wise man. During slow times, they played gin rummy together and just chatted about world events. During World War II, my father and Doctor Shapiro had a noteworthy discussion about the infectious discrimination and destruction that both of their ethnic populations were facing. Japanese were concentrated into camps in the American Midwest, while Jewish people were forced into work and death camps during the Holocaust. Doctor Shapiro remarked, "Jim, what your people are experiencing now will subside after the War, but what my people have and continue to experience will never end."

I believe that the Jewish and Japanese Americans have a special bond because of the shared experience of prejudice which we have endured. But there is a deeper connection, and that is a strong belief in the values of our cultures, especially of family and education. My parents always stressed to us that education is one thing that can never be taken away from you. Both Jewish and Japanese Americans learned this firsthand during the internment camps of World War II. I also saw these paralleled values with our dentist, Dr. Seymour Appell, who was also Jewish. His practice was right near our gift shop, and his legacy was carried on by his three sons, who all became dentists as well. I still go to their family practice, and I joke with his son, Joel Appell, that one of their patient rooms should be named the Mukoyama room. Through their education and family values, the Appell Dental Practice has maintained dental professionalism and a caring practice.

In my high school years, I was saving up to fund my college education through playing in a band. My friend Ronald Simon and I joined together to play weddings, bar mitzvahs, bat mitzvahs, and dances. We became very familiar with Jewish temples and Catholic parishes through the city and northern suburbs. Through these musical ventures, I gained an understanding of different religious congregations and an appreciation for the education I was saving for. Every step of my upbringing was guiding me towards my future, but there was one lesson that I could only learn in the world's best-selling, most well-read, and historically influential book.

My most important connection with the Jewish people has been the Holy Bible. The Christian Bible is comprised of 66 books, with 39 books making up the Old Testament, which are actually the Jewish canon from creation until around the early 400's B.C.; and the 27 books of the New Testament, which cover from the birth of Jesus to his death, resurrection, and the book of Revelation, written by the disciple John in the last decade of the first century A.D. The first five books of the Old Testament comprise the Torah and are the basis of the Judeo-Christian faith.

It is the Old Testament that Jesus and his disciples quoted and therefore is the foundation upon which the Christian religion stands.

I became increasingly appreciative of this special relationship through other Jewish people who have entered my life that you will hear about in later chapters.

I was not always as knowledgeable about the Scriptures as I am now. Although I have had my own Bible since I was promoted from the primary to the junior department of the Avondale Methodist Church School in 1953, I had not read it from beginning to end. I began pouring over the Bible when I was in college, and I've been studying it ever since.

I still have that Bible with the cover I made at the Moody Day Camp. Nearly 60 years later, KJ and I were at an engagement party for our daughter Sumi and then-fiancé John Ekholm, hosted by his sister, Anna Jones. This was the first time we had met most of John's extended family members. I noticed a gentleman sitting by himself, so I approached him and introduced myself. His name was Paul, and I discovered he was John's uncle, a minister, and from Chicago. He said his father was the pastor of a Methodist church in Logan Square. When I asked him his last name, he said, "White," to which I responded, "William D. White." His father was the man who signed my Bible over half a century earlier. I considered this incident a blessing from God on the coming marriage.

As the years went by, I participated in numerous Bible classes and studies, including a two-year deep dive into first the Old Testament and then the New Testament. I learned that the Bible begins, continues throughout, and ends with the theme of love.[1] At the end of 2020, I gave my wife a gift of a three-ring binder with an overall history and organization of the Bible with a description

---

1   Quote from *The Greatest Love Letter Ever Written*, by James H. Mukoyama, Jr. on January 3, 2021. The essay can be found in Appendix A.

of each of the 66 books in detail including the author, time written, and main theme.

On the first Sunday of 2021, I got an inspiration to write an essay about the Bible, entitled, *The Greatest Love Letter Ever Written.* One does not normally qualify an anthology as a letter, but the Holy Bible was indeed a love letter. The author is the Creator of the universe, God, and the recipient is anyone who has read it, in whole or in part, in the past, present, or future.[2] I had not planned to write the essay — the words just flowed out of me. I completed my writing in less than 40 minutes, and I have not changed a single word. It is one of the few times in my life when I can say that I know my actions were completely controlled by the Holy Spirit, another "God Thing." The Bible, as any love letter, needs to be cherished and enjoyed over and over, for the more you read it, the more you learn and enjoy. I continue to revisit this glorious writing, because the author has written it and created this universe, world, and humanity as an act of love.

As I came of age, these Judeo-Christian moments and experiences shaped my perspective and beliefs. The "God Things" had been coming and going throughout the years, but soon enough, I had to decipher what path God envisioned for me.

---

2   Quote from *The Greatest Love Letter Ever Written*, by James H. Mukoyama, Jr. on January 3, 2021. The essay can be found in Appendix A.

CHAPTER 5
CROSS OR RIFLE DILEMMA: ANSWER TO PRAYER?

I have always sought opportunities to serve, but in high school, I encountered a crossroads. I'd been inspired by the example my parents had set, supporting and serving others, especially amidst the difficulties of wartime. But I was torn between two callings — faith or the armed forces.

As I explored different Christian denominations, I felt a pull onto God's path. At the invitation of my second cousin Martha Hikawa and her husband Gus, I joined a youth group at their church, Christ Congregational Church, or Tri-C. Their Congregational denomination church was an ethnic Japanese American congregation with two services, one in Japanese for the issei immigrants, and one in English for those who were born here, many of whom did not understand Japanese. For the first time, I met fellow teenagers of Japanese ancestry, and I was moved by my connections with other young people, my ancestry, and God himself. For the first time, I met teenagers of Japanese ancestry from all over the Chicago area. While my immediate family continued to attend our previous congregation, I gained independence by joining Tri-C on my own. I immediately became actively involved in the youth program called Pilgrim Fellowship, or PF, at Tri-C. Each meeting had a worship service, community or church service activity, and some fun activity, such as pizza, games or dancing — both up-tempo and slow under the watchful eyes of our ever-present advisors. PF was a very effective organization with elected officers: President, Vice-President, Secretary and Treasurer. It had three committees with chairpersons for each: Faith, Action, and Fellowship.

From an early age, I ventured into leadership opportunities in both my school and church. When I was at Avondale Elementary School, I ran for office in the student council three times. Although I lost in my first two attempts, the third time was a charm, and in my last year, I became President of the student council. Because of this experience, I had meaningful contributions with which to support the Pilgrim Fellowship. After members of PF got to know me, I served as Chair of the Action Committee, then I was elected as President. I then extended my activity to the entire Chicago Area of Pilgrim Fellowship, as I became Vice-President of the overarching organization that included dozens of churches throughout the Chicago Metropolitan Area. I was on fire for my faith and was praying to enter the ministry after high school.

Yet, my samurai instincts were starting to kick in, and my hobbies were aligning with the military. The samurai cultural influence cannot be understated, as I looked up to revered warriors in my Japanese ancestry. The famous samurais, Takeda Shingen and Saigo Takamori, were two prominent warriors from my father and mother's family prefecture. In our family home, etchings of them hung proudly on the wall. Following in the footsteps of my older brother, I became active in my high school Junior Reserve Officers Training Corps Program, or JROTC. Because of my activities in the band, I was an officer in the JROTC band. Our JROTC training was strictly in Infantry, and we had a 22-caliber rifle range in our school basement where we learned to shoot. My former Boy Scouts training also made it easy for me to be disciplined and focused, and to follow and lead. My success in these arenas and my samurai inclination had informed my life, and now I found myself torn by two passions.

I had a dilemma. I wanted to serve God, but I also wanted to serve in our military. Both were incredible opportunities to serve, but completely different routes.

At the same time, our Tri-C pastor was getting older and looking forward to moving back to the West coast. The church elders knew about my passion for our faith and desire to become a minister, so they offered to pay for my college tuition and seminary. This offer was incredibly tempting. I had worked several jobs starting in elementary school, so I understood the value of a dollar and the investment the church was making in me. After seminary and ordination, I planned to return to our church and assist the pastor until he decided to retire. I know God has been with me my entire life, but I can only recall a few specific incidents where the Holy Spirit intervened. This moment was one of them.

As tempting as the offer was, if I accepted the church's generosity, I would then be fully committed. If I did not get ordained for any reason, I would be letting everyone down and not fulfilling my word. So, after prayerful consideration and guidance from the Holy Spirit, I politely declined their generous offer with gratitude. But I still was determined to enter the clergy.

I came up with a brief solution that solved my dilemma by satisfying my desire to serve both my passions: I could become a chaplain in the Army. I would go to the university and while getting my degree, go through the senior ROTC program and get commissioned as an Infantry Officer, as my samurai instincts called for again. Then, after I went to seminary, I would return and transfer from the Infantry to the Chaplain Corps. In that manner, I concluded that I could be a more effective shepherd to my flock, having gone through the same training and frankly, earned some street cred with my soldiers.

But then, everything changed. My Congregational Church denomination merged with the Evangelical & Reformed Church denomination to form the United Church of Christ. The new denomination went far to the left theologically, and I could not accept some of their tenets. My faith was still rock solid, but I disagreed with their policies. This put me in a Catch-22 untenable position. In order to become an Army Chaplain, my denomination would have to endorse me, but in good conscience, I could not accept their theology. Through my prayers at the time, I concluded that God wanted me to concentrate on my military career. I continued to be a member of the church but could not pursue my chaplaincy under false pretenses. I knew that God had a plan for me, so I followed His path.

From that point forward, I devoted all of my energy to becoming the best Infantry Officer I could be. I attended the University of Illinois and participated in everything I could that involved the military. If it looked, moved, or smelled like military, I was in. I joined ROTC again, which molded me into a disciplined, hardworking man. I pledged for a military fraternity, Pershing Rifles, and joined the drill and rifle team. Every year the Chicago Tribune awarded gold, silver, and bronze medals to university cadets in each grade group. I received the Gold Medal as a junior cadet. I became an officer in Pershing Rifles, and I also was a member of another military honorary society, Scabbard & Blade, and became the commanding officer. I then attended our senior ROTC Active-Duty field training between my junior and senior years at Fort Riley in Kansas. Out of over 2,000 cadets from all over the country, I received the 7th-highest rating, which was the highest for any cadet from my university. For the mid-camp review, I was selected to be the Commander of Troops. My decision, to follow God's plan into the military, was validated by my undeniable success in these endeavors.

ROTC Summer Training. Fort Riley, KS. 1965

However, the next year, I was passed over to be appointed as the Cadet Brigade Commander. Another engineering student was chosen, and it turned out his father was in charge of the Chicago high school JROTC. I did not victimize myself or play the race card. But 20 years later, I visited the retired secretary of the ROTC Department, Mrs. Staley, to thank her for her motherly guidance of myself and the other cadets. She disclosed to me that I clearly should have gotten the position. I was instead offered another high-up position, but I realized my passion was in leading fellow men. I requested to become the Commander of one of the battalions, and it was granted. This experience helped me later in my career and became a pattern that held true. I always preferred command with troops versus a desk job.

As college graduation approached, I applied for a commission as a Regular Army, or RA, officer in the Infantry branch. Since Vietnam was heating up in 1965, the branch assignment was nearly automatic. There are two types of Army commissions, Regular Army (RA) and Reserve. Only a highly select few ROTC graduates receive RA commissions, because Regular Army officers receive benefits that support a life-long Army career, no matter if the country is at war or at peace. When the Army is decreasing its numbers during times of reduction of force, the remaining Reserve soldiers' rank is reduced since there are few soldiers to command. For example, the Army was dramatically reduced after World War II, and some Colonels were marked down to Sergeant. But Regular Army officers cannot be demoted lower than Second Lieutenant. Most RA officers' ranks are not reduced very much, if at all, and preferences for the best assignments often flow to them. An RA officer designation marks you as a "lifer," or career officer.

I was on my way to receiving an RA commission, but I had to get two waivers. First, there was a height minimum of 5 feet 5 inches, and I was only 5 feet 4 and ½ inches. I'm proud of that ½ inch, but I had to be waived to account for the other missing half. Second, I had to get a doctor's letter vouching that my past operation to remove the tumor on my head would not be a physical impairment in combat. So, I went to see my life-long family physician, Dr. Shapiro. At the time, Vietnam protests nationwide were ensuing, with guys burning their draft cards and booking for Canada to avoid service. I walked into Doctor Shapiro's office and said, "Doc, I need a favor from you regarding the Army." Before I could get another word out, he responded, "Don't worry Jim. I'll make sure you don't have to serve a day." Needless to say, I immediately corrected that misunderstanding, and he wrote the waiver justification. I received my Regular Army officer commission in the Infantry branch when I graduated in June. My Army career was on its way. I concluded the answer to my dilemma — ministry may be a "no," but the military was a hard "yes."

CHAPTER 6
REACHING FOR THE STARS: WHICH ONES?

Above the entrance to the library of my high school, there was a quote from Carl Schurz: "Ideals are like the stars; you guide your life by them, but never reach them." What keeps you from reaching those ideals?

Let's look at a guy from the inner city of Chicago. He starts out life in a lower-middle-class, blue-collar neighborhood, in a two-bedroom apartment on the third floor. His father is an immigrant of a very small minority population in the area, with no college education, professional training, or special skills. His mother has no university experience, nor has she earned a paycheck, and she was unfamiliar with this geographic area of the country. Her parents live with his family, and they are both immigrants that barely speak English and have no income. Although their family and community are close knit and conscientious, they are hindered by the challenges of their situation.

That's for starters. He is often overlooked based on his small size and outside appearance. In terms of physical stature, he is always the smallest in his group wherever he goes — school, church, or just outside to play. He strives to advance ahead of his peers, but he is also always the youngest in his group cohort. And finally, he looks different from everyone else because of his race and a scar on his face.

This reality is what I experienced in my early life, but it was *not* a recipe for failure in life. It was just the opposite. I always saw and understood life not as filled with obstacles to overcome, but rather opportunities to excel. There is a phrase in Japanese, *Shi gata ga nai*, which means "It can't be helped" or "It is what it is." I neither expected nor was I promised that life was going to be a rose garden. I had to plant and cultivate the seeds of opportunity if I wanted my future to blossom.

Because I was so small, I did not excel in the normal sports of baseball, football, track or basketball. In neighborhood baseball games, I was always chosen last and put in right field, the Siberia of baseball fielding positions where you can do the least harm. Though the neighborhood kids never said it aloud, I was smart enough to realize that the decision was totally driven by my appearance. I was actually an excellent fielder and a pretty good hitter, not for power but for average. When I was in eighth grade, Rogers Hornsby, the Chicago Cubs Hall of Fame player came to our school, Avondale Elementary, and watched us play. Since I was determined to show my skills, I went five for five and made no errors in the field during that game.

In the early grades of grammar school, I was also a target for bullies during recess. But these kid bullies didn't know that I've always had a bad temper. I hid this fury in school because I never wanted to get in any trouble, not because I was afraid. My parents had stressed for me to behave in school and study hard. The last thing I ever wanted was to bring a note home from my teacher, or worse, the horror of horrors, to have my parents called to the school for a meeting with the principal. I feared the consequences of the discipline I would receive from my parents more than from any school official. I realized that if I did not control my temper, I would face my worst nightmare — disappointing and disrespecting my parents. So in the early school years, I gritted my teeth and ignored the taunts, telling myself to consider the source as not worthy of response. But when I was 8 or 9, I lost it. The school bully called me a "Jap." Something in me just snapped, and within seconds, I had him on the ground and was on top of him beating him. Eventually, I was pulled off by my classmates. They stood there in a state of shock. What happened to the Jim Mukoyama that they knew? I didn't break anything, and the bully was more embarrassed than hurt. But this uncharacteristic outburst left an impression on the school bullies, my neighborhood friends, and even myself.

From that moment forward, I was left alone. Bullies knew that I was no longer someone they could mess with. Most importantly, nobody ratted on me. The kids honored the neighborhood code and handled matters privately, and I had no letter to take home. I never mentioned that incident to my parents. I got lucky, and I learned my lesson.

Since I was also the youngest in my peer group, I often struggled to gain proper social skills. Being the youngest in my peer group growing up was harder to overcome because I often lacked the social skills developed in peer experiences. Charisma and confidence often developed through interpersonal relationships and peer experiences, but in my youth, I lagged behind. When high school started and testosterone levels began to kick in, my disparity became even more apparent, especially with girls. To say a "God thing" happened in this area might seem to be sacrilegious to some, but the acceptance of one girl truly helped shape me into the man I am today.

Here's what happened. When I was a junior, to say my dating

life was nil would be an understatement. I did hang out with a great group of guys and girls who were in the band and orchestra, but not being a jock and not having a car, my odds of attracting any girl were slim and next to none. There was a very beautiful blond-haired, blue-eyed girl named Ginny that all the guys admired. One day, the girlfriend of one of my best friends asked me, "Jim, why have you not asked her out on a date?"

Honorale Schurzites - High School Junior Class Service Award. Schurz High School. 1961.

I thought she was kidding. First of all, the girl was a whole three inches taller than I, and I wasn't exactly a great catch. But my friend said I might be surprised at her answer. Now, I might be thick-headed, but I didn't need any further encouragement. I asked the girl out. To my pleasant surprise, she accepted. We went steady throughout high school, and she even started wearing my high school ring. She eventually dumped me, not for anything I had done or failed to do, but from that point on in life, I never doubted my ability to establish a relationship with a girl. Regardless of my circumstances or the odds of success, I was confident in myself.

Rather than assessing based on appearance, I proved to myself and my classmates that you cannot judge a book by its cover. As one of six students of Asian descent, I stood out as different, but due to my small physical stature, I was often overlooked. Rather than harping on these disadvantages, I looked for opportunities to get involved, improve, and excel. In the school band, I worked my way up from beginner to Symphonic Band through my dedication to the clarinet. A pattern was emerging within myself. Once I involve myself in a project or passion, I'm all in.

In college, I took up martial arts, specifically karate. Not to break boards, but to gain self-control. While I knew I had to restrain my temper, I wanted to master it. I was fortunate to study under a black belt sensei (teacher) qualified by the Japan Karate Association, or JKA. This was a serious classical program that emphasized form, repetition, concentration, and meditation. These core tenets instilled self-discipline, and the results have been extremely important and lasted a lifetime. Martial arts can give you a personal calmness. Once you reach a certain level of proficiency in the martial arts, you understand that your body, if uncontrolled, can become a lethal weapon. You must be aware of these capabilities if you get into a dangerous situation. But because I know that I can do serious harm in dangerous circumstances, I don't have to do anything to fight back. I know I can win.

Later in life, I renewed my martial arts pursuits when our young son Jae and I both enrolled in a local Tae Kwon Do program. It was a great father/son experience. Jae received his black belt before he was 10 years old. I attained my brown belt but sustained a leg injury that ended my martial arts career in my 40s. On the other hand, Jae has continued his martial arts pursuits and qualified as a certified personal physical fitness trainer. I was proud to watch him follow his own directional stars. Although we were destined for different life paths, we followed similar ideals.

My directional stars in my life pointed me in the direction of becoming a General in the Army. The odds of that becoming reality at the time were implausible. There had never been a General officer of Asian ancestry up to that point. But I didn't let how far-fetched this starry ideal was stop me from reaching for it.

When I was a young Lieutenant on active duty, I received advice on things I should do or learn if I wanted to become a General. Folks said I should learn how to play golf and bridge because these activities would become opportunities to network and make friends with senior officers. But they were expensive and time-consuming, and ultimately, I saw no connection between these skills and leading soldiers. As a result, I never did learn to play golf, and although I was an excellent card player, I never pursued bridge, nor did I suggest that my wife learn.

What was more important to my career as an officer were the great non-commissioned officers (NCOs), the sergeants, who made me look good, and the senior officer mentors who didn't cut my head off when I made a mistake. I made my fair share of mistakes, but I had leaders who humbled me and subordinates who believed in me. This combination of encouragement and humility kept me balanced yet determined.

One such NCO that I had the absolute honor of commanding stands out, Allen J. Lynch, a Medal of Honor recipient. He was one of my company First Sergeants when I commanded a battalion of the 85th Division (Training) in Waukegan, Illinois. While serving in

Vietnam as a radio telephone operator in the 1st Cavalry Division, Al rescued three of his comrades and carried them to safety under intense enemy fire, then single-handedly defended their isolated position for hours. I was incredibly blessed to have a living war hero in one of my key leadership positions.

Through leading by example, Allen Lynch uplifted my command through demonstrations utilizing his insight and experience. His leadership in the 85th Division was crucial because we were converting from an Infantry training division to an armor/cavalry training division. Having served in the 1st Cavalry Division in Vietnam, he fully understood the equipment and objectives that we were working with. One moment we were training 11B ("eleven bravos") MOS (military occupational specialty), or Infantry soldiers, and the next we were training 19D ("nineteen deltas") MOS, or cavalry soldiers. We were teaching our recruits how to operate and maintain M113 Armored Personnel Carriers (APCs) for their cavalry scouting missions. Allen had worked with APCs firsthand in combat, and our soldiers knew that. I leaned on him too, and he came through. At Fort McCoy in Wisconsin, we needed to convince former Infantry soldiers that the 12-ton tracked vehicle APCs could float and move in water. There was a small lake at Fort McCoy, so we positioned our skeptical soldiers in bleachers on one side. The only way to overcome the fears and doubts was to demonstrate the task could be done. On the other side, I commanded an APC that entered the water, moved across the lake, and came out of the water in front of the bleachers on the other side. I got outside of the track and announced, "See, it works! Now it's your turn."

So, the foundation of my life was poured. Often, I had to redefine perceptions of myself, obstacles that stood in front of me, and directions I was being pulled in. But once I had overcome these challenges, I saw them as opportunities that catapulted me closer to reaching my directional stars. The next section will address critical life experiences moving forward and how those foundations were called upon.

EXPERIENCE

CHAPTER 7

M*A*S*H* COMES TO LIFE: MY ARMY CAREER
AND KOREAN CONNECTION STARTS

In November 1966, I was an Infantry Second Lieutenant, ready to fly overseas the next morning to join the 2nd Infantry Division on the demilitarized zone (DMZ) in the Republic of Korea. The Infantry is the elite Army branch that specializes in ground combat on foot, and I was ready and enthusiastic to join their ranks. I was at a hotel near the Seattle-Tacoma International Airport, eagerly anticipating the beginning of my career as an Infantry Officer.

I received my master's degree in the Teaching of Social Studies the previous June. When I received my bachelor's degree the year prior, I was offered a Resident Assistantship to be a counselor in the dorms. It was a wonderful opportunity to receive my graduate degree with all tuition and books paid for and a salary of $100 a month. Upon receiving my bachelor's degree, I was also commissioned as a Regular Army officer and in order to attend graduate school, I had to take a one-year unpaid period, called excess leave. Luckily, the Army was more than willing to grant me the leave because I would be a more qualified officer.

The year of 1965 turned out to be a mixed bag of ups and downs for me. On the upside, I was able to continue my personal research into my East Asian background. Since I had made a decision early in my career to become an Army Infantry Officer, I was able to pick any academic major I wanted. I narrowed my selection to two subjects, between Psychology and English literature. Although math was my best subject and history was my favorite, Psychology and English were chances to study and better understand people and relationships, which would benefit my Army career and leadership. After my first semester of studying Psychology, I rejected it after studying Pavlov and his dogs. I failed to see how the subject connected to working and interacting with people every day.

On the other hand, English literature, with its full continuum from Shakespeare to Hemingway, revealed common traits of human nature. While reading the classics, these observations jumped out at me. If you can grasp these concepts, you can better understand what motivates people. Furthermore, by reading the greatest writers of all time, I figured that some of their talent may rub off on me just by sheer osmosis, and I would be able to express myself better. So I elected to major in English literature, with a minor in history, and it proved to work out for me in the long run. But in the short run, since I was not a speed reader, I spent many late nights reading at the undergraduate library and in my dorm room.

Speaking of my dorm room, I couldn't afford to live in an off-campus apartment, so I lived in the dorms in a six-building complex referred to as "six pack." My roommate at the time was a fellow Pershing Rifleman from Chicago whom I will call Frank. Being the all-American healthy male that I was, I was a reader of Playboy magazine and had kept a collection of the "artistic" centerfolds. I had saved so many that I papered the ceiling of our dorm room and one wall with these pictures. The word got around campus, and we had guys come to our room just to see our décor. One weekend, my roommate's mother visited campus and showed up unexpectedly. When she entered the room, she sighed and looked at her son and said, "Oh, Frank!" I immediately looked at Frank and said, "Yeah, Frank!" I bet some of my friends reading this didn't know I was such an art aficionado.

During my junior undergraduate year, another "God thing" occurred that became the critical puzzle piece to connect me to my ancestry and career. The University decided to expand its Asian studies program, specifically adding professors in Japanese history, political science, and language. Prior to that, the Asian studies program offered a robust curriculum regarding China, but not Japan. I immediately signed up for every class and subject regarding Japan. I did the same for Chinese history and political science because to understand East Asian countries, one must understand China. Its language and philosophies were the foundation of all the cultures in that region of the world. My understanding of Asia proved to be especially crucial in so many ways throughout the rest of my life, but especially in the next five years.

On the downside, my personal life hit a landmine. A girl, to whom I was virtually engaged, sent me an unexpected "Dear John" letter just after I joined the military. She came from another Japanese American family who had very close ties to mine. Her father and my mother knew each other in their early youth in Madison, Wisconsin, and we used to visit them during the summer at their small farm where they grew vegetables. Her father was a World War II Veteran and a member of the famous 100th Battalion/442nd Regimental Combat Team that served with exceptional valor in the European Theater of Operations (ETO). He was wounded seriously in France and was very active in the Military Order of the Purple Heart throughout his life. He had three children, all girls, so I tried to absorb all the insight he would have passed along to a son. I was able to get as much advice from him as I could by sharing our common military interest and experiences. He was a man whom I admired greatly, and I felt extremely fortunate and honored to have known him. So, I was crushed and frankly couldn't understand what happened. But there is that Japanese phrase which covers most life unpleasant situations, *Shi gata ga nai* meaning "it can't be helped" or "it is what it is." More importantly, I have read the Bible. I know how the book ends, and I know that God always has my best interest in mind, no matter how tough things get. So, I sucked it up and moved on, thinking, "This too shall pass." Rather than wallowing in my heartbreak, I focused on my pending deployment and Army career.

Now in Seattle, I contemplated the job that was awaiting me across the Pacific. I was really looking forward to this new beginning, but alone in my hotel room, I began to brood and get a bit homesick. A good friend of mine, Fred Mann, who went through Infantry Officer basic and Airborne school with me, was from Tacoma, Washington, about 30-40 minutes south of Seattle. Fred was already assigned to the 2nd Division near the DMZ. I had his parents' phone number, so I gave them a ring.

"Hello Mr. Mann. I will be seeing Fred soon when I deploy. Is there any message I could convey for you?" I asked.

Fred's father, Mr. Harold Mann, was an insurance executive and a strong man of faith, well known in his community for his professional and charitable work. My tone of voice must have indicated I was lonely because he immediately offered to come to my hotel and take me to dinner. He took me to an exclusive restaurant in downtown Seattle, and when we sat down to eat, Mr. Mann said grace, thanking the Lord for all His blessings and asking for His protection of our service members. I was not accustomed to saying grace before every meal back then, just on special occasions, so this practice caught me by surprise. But Mr. Mann's gratitude inspired me, and since then, I've consistently incorporated saying grace into my own life. The Mann family set an example on how I could live a more virtuous, holier Christian life. In the next four years, the Mann family was instrumental in my becoming more connected to my Christian walk — another "God thing."

The next morning, I departed McChord Air Force Base and flew to Kimpo Air Force Base in South Korea. When I stepped off the plane and into Seoul for a short time, it was the first time in my life that I didn't feel like a minority — everyone was short, with black hair and brown-eyes. It was an unexpected but welcomed experience.

One of my Pershing Rifle fraternity brothers had served in Korea and told me to say hello to Miss Kim. And the first Korean woman I met at the airport snack bar, lo and behold, had a nametag with "Kim" on it. I was amazed. I thought to myself, how did he know I would meet her? But then I met another Miss Kim — and another. I soon learned that Kim is among the most common surnames in Korea, along with Park and Lee.

In Korea, I gained first-hand experience through practical conflict and leadership examples, while also being blessed with a comprehensive understanding of Korea. I reported to Company A, 1st Battalion, 23rd Infantry of the 2nd Infantry, or Indianhead Division, which was located right outside of the DMZ, north of the Imjin River and northwest of the capitol, Seoul. As an RA Infantry Airborne Officer, I had volunteered for Vietnam but was sent to Korea instead. I was initially disappointed, but this experience turned out to be a true blessing, or "God thing". Not only was I able to lead Infantry soldiers in combat patrolling, but we also had our own company compound, Camp Young. Our compound was surrounded by barbed wire and required a daily armed guard mount, and we performed all of the routine administrative functions, such as Command Material & Maintenance Inspections (CMMIs) and Inspector General (IG) Inspections, which are normally reserved for non-combat garrison duty. Camp Young was virtually the same as the 4077th Military Hospital in the TV series M*A*S*H, complete with a helipad. Despite the need for Infantry Captains in Vietnam, our company also had exceptional leadership, as it was commanded by a Captain who was a West Point grad and Vietnam Veteran. Each of our four platoon leaders were Regular Army Airborne officers, so our company leadership was rock solid.

The coldest day in my life was etched into my mind as a clear memory and critical leadership lesson. It was Christmas Eve 1966, and I was leading a combat ambush patrol within the DMZ. The DMZ was considered a free fire zone, and no one was authorized to be in the DMZ except our soldiers. This meant we were free to open fire on anyone detected, no questions asked. We walked out into our positions at dusk, and we had to remain silent and motionless until daybreak the next morning. This was no easy task. Korean winters are extremely frigid, with temperatures averaging a low of -4 degrees Celsius, or 27 degrees Fahrenheit. We dressed in layers and carried dry socks, which we had to change into once we got to our position because we sweat just by walking to the position from our camp. Being the patrol leader, I felt grateful for the luxury of moving to check my men's positions throughout the night. But since I was leading twelve soldiers, and even one platoon sergeant who survived the Korean War, I had to be a humble example and could not complain for even a second. I was shivering from the cold but couldn't let on to my troops. As the leader, I had to grin and bear it, because my resilience motivated my troops to stay strong too. Truth be told, that was one of the longest, most humbling nights in my life.

2nd Division Staff Officer. DMZ, Korea. 1966

After about four months, I was ordered to report to the Battalion Commander, LTC Dewitt Cook, who was a Korean War Veteran with a Combat Infantryman Badge (CIB), the most valued military award for an Army soldier. Of course, I am slightly prejudiced as an Infantryman. LTC Cook was one of the finest commanders that I encountered in my entire military career. Although I was not awarded a CIB during my time in Korea, I did earn an Expert Infantryman Badge (EIB). The EIB involves a series of numerous tests, mostly in the field, of Infantry skills, such as marksmanship, first aid, land navigation, road marches, physical tests, explosives, and radio communications. Infantry divisions administer these tests on a regular basis, and the qualification rate is less than 5%. For example, when I took the test, there were only about 40 of us who qualified for the badge, out of our whole division of approximately 13,000 soldiers. My small T-shirt swelled to medium when I saw that I attained the highest score in the entire division.

Reporting to the Battalion Commander was a very rare occurrence, so I was nervous yet intrigued. Junior officers typically only meet with Battalion Commanders when they're in trouble, so I didn't know what to expect when I reported to LTC Cook.

"How are you doing?" LTC Cook asked me when I met with him.

"I am finally leading Infantry soldiers, after eleven years of preparation starting in high school JROTC. Our soldiers and company are outstanding," I responded eagerly.

He looked down at my paperwork and noticed I had a degree in English and a master's degree. "How would you like to be the Battalion S-1, or personnel officer?" He asked, offering what is normally a Captain's position.

"I am honored to be even considered, but I'm very happy where I am and would prefer to remain on the line," I answered.

"I'm not looking for happiness in the Battalion. You are to report the following day for duty as the Battalion S-1," he replied sternly.

I had experience serving as a personnel officer from my college military fraternity, but my greatest passion was serving and leading Infantry Officers. I was a stickler for knowing and following Army Regulations, but I also knew how to deal with the company commanders as a staff officer. But as a staff officer, I learned skills about how to interact in organizations. My main function was to serve the commander, advising and assisting him on personnel and administrative decisions. I had to become an expert in my department and become familiar with other staff members' areas of expertise, so that I could build relationships and coordinate with them. Through this critical learning experience, I came to understand how to work within and interact with different departments of a larger organization.

When the Brigade Adjutant position opened up several months later, the Brigade Commander, with LTC Cook's recommendation, selected me, a newly promoted 1st Lieutenant, to fill that vacancy, normally a Major's position. These staff assignments furthered my career and helped me serve my Infantrymen, but also positioned me to make a difference in the lives of the Korean people.

CHAPTER 8
HONORARY KOREAN CITIZEN: MAYOR OF CHANG-PA-RI?

As an Asian American with a global awareness and an inclination to serve others, I went to Korea with some preconceived notions and devotions. When I was a student at the University of Illinois, *The Ugly American* was a very popular book about how Americans' behavior while traveling overseas had caused very unfavorable feelings toward the United States. In essence, American tourists would expect people to speak English, and often acted in a condescending manner to those who didn't. I vowed to myself that I would do all I could to reverse that trend when I went abroad. This attitude was not rocket science — I just treated everyone I met with respect as fellow humans. I always tried to remember that I was a guest in their country, so I treated them like I would want tourists to treat me in America. In fact, I found it to be helpful to learn about the country's customs, values, and language before visiting. With this mindset, I entered into South Korea cautiously, not knowing that I would end up forming an invaluable, life-long relationship with this land and its people.

The U.S. 2nd Infantry Division was stationed in an area of South Korea that included a portion of the Demilitarized Zone which we were responsible to defend, including the U.N./North Korean negotiating area of Panmunjom. Our brigade area of operations included the Imjin River in a Province called Pagu-gun. Because of its location near the DMZ, Pagu-gun was one of the poorest parts of the country. The central government did not invest heavily in the infrastructure because it would be the first to go in time of war. There were no paved roads, only gravel. There was very little running water, and electricity was limited to a few hours in the evening. Pagu-gun was an extremely impoverished rural area, and for many of us American soldiers, it opened our eyes to these difficult living conditions.

Not only was the area affected by its proximity to the DMZ, but the Imjin River shaped the territory and its village too. The Imjin River is a very wide and deep river, and there were only two large bridges able to cross it in our division sector of the DMZ. The river made a large half-U-shaped turn that looked like a spoon or a duck's bill, so consequently, the region was nicknamed the "spoon bill" area. The Libbey Bridge was situated in the northeast, and the Freedom Bridge was near Panmunjom in the southwest. My brigade was responsible for the Libbey Bridge sector. Prior to crossing Libbey Bridge into the DMZ area, there was a small village called Chang-Pa-Ri, but colloquially referred to as "the Vill" by soldiers. Chang-Pa-Ri had a few shops and a large bar & grill, where the soldiers, or G.I.s, would hang out on pass in the evenings and weekends when they were not on duty.

Although Korean and Japanese are deemed contextually similar in their migration history and physical appearance, these two groups have a long and strained history. At this point, it is necessary for me to add a bit of historical context on the relationship between Japan and Korea. I never studied anthropology, but the Korean and

Japanese people are among the closest in physical appearance than any other two major East Asian people. I realize that most Caucasians perceive that Asians — Chinese, Korean, Japanese, Vietnamese, and others — all look the same. But I can assure you that we of Asian ethnicity can distinguish between those different groups fairly easily. Not only do the Koreans and Japanese have similar physical characteristics, but it would also seem that the population movement began in ancient China, moved through Korea, and ended in Japan. But throughout history, Korea and Japan developed into two very distinct populations and nations with occasional unsuccessful military incursions against one another.

As a result of the rise of the Japanese Empire in the late 19th century via victories in the Sino-Japanese and then Russo-Japanese Wars, the Empire of Japan annexed the Empire of Korea in 1910 and occupied Korea until the end of World War II in 1945. Under a ruthless policy of "Japanization", this occupation entailed total immersion in the Japanese language and culture. Accordingly, Japanese became the official language taught in Korean schools; Korean names were changed to Japanese pronunciations; and Korean patriotic songs were banned. Violations of the Japanization policy resulted in harsh punishments, ranging from exile to execution. Koreans' animosity intensified against the Japanese, and their relationship was riddled with distrust and hatred. To this day, Korean people still celebrate the day World War II ended and Japan lost. Since I studied Japanese history, I was well aware of the environment I was entering. Quite frankly, I was very concerned about how the Korean people would accept me, an American of Japanese Ancestry. But I should not have worried.

Throughout my entire 13-month tour in Korea, the people were welcoming, kind, and grateful for America's sacrifice of over

37,000 United States Armed Forces members, who died defending their country against Communism. The elderly especially welcomed me. They wanted to converse with me in Japanese, and fortunately I had studied Japanese language, so I could at least carry on light conversation. The fact that I was of Asian descent was actually a benefit, Japanese or not. By treating them with simple common courtesy and respect, I developed relationships based on mutual respect and trust.

Chang-Pa-Ri residents were extremely poor, and many community social services were lacking. Through my conversations with the Korean populace, I understood how dire their needs were. As the Battalion Adjutant who was responsible for unit administration and resources, I saw an opportunity to fill the gap in the local community. The community had countless needs, so how could we make a difference?

I established the Chang-Pa-Ri Community Assistance Fund to affect a broad base of change. Our fund was going to be run by non-commissioned (NCO) officers, so I appointed the Battalion Command Sergeant Major to be chairman and the First Sergeants to be directors. I was just the man putting it all together, but luckily, I chose the correct leadership to be on the board. By selecting senior NCOs, soldiers were compelled by their elder peers, rather than feeling forced by their superiors, and were confident in the projects the money was going towards. Records of any monies collected were strictly reviewed by the board, and all decisions for expenditures were likewise decided by them. Accountability and transparency were in order, so soldiers knew this fund would be aiding the community of Chang-Pa-Ri.

In those days, a soldier's pay was distributed monthly in cash and in person, minus any check allotments to family. In my case,

I sent a monthly portion of my salary to support my parents. On payday, the pay officer would have a ton of cash, his pistol, and an armed guard. Being in a combat zone, instead of U.S. dollars, we were paid in Military Pay Certificates, or MPCs, that could be used in the local community or exchanged for local currency. Soldiers would stand in line and wait for their income, and the pay officer would sit at the head of a long table with his records to pay each soldier. Local vendors, such as the barber, cleaner, and tailor, would also be seated at the table, to collect their tabs. At the very end of the table, I placed a coffee can simply labeled "Chang-Pa-Ri Community Assistance Fund."

Americans are generous people, and our soldiers were no exception. With their pockets flush with cash, they would drop 25 or 50 cents into the can. With a battalion of 1,300 soldiers, a monthly collection would be a significant contribution for the fund. On average, we would collect nearly $400, which went a long way in this small village. Through the fund's bottom-up organization, soldiers could be proud of the impact they were making in Chang-Pa-Ri, both on and off duty.

The board decided that the fund should be dedicated to building up and assisting the community, especially to support families and aid in emergencies. The village had two schools. While there was a private school for the rich, the public school had mostly volunteer instructors and meager resources — no desks, no chairs, and no blackboards. After the war, wood in Korea was scarce and expensive, but our fund managed to purchase twenty desks, forty chairs, and three blackboards. Then, three of the village homes were destroyed in a fire. We gave each family cash grants to help them rebuild. The local village Presbyterian church had wooden doors and stairs at the entrance in need of repair. As the Battalion Adjutant, I had access to all of the personnel records of my soldiers. I was able to determine soldiers in our unit who were carpenters in civilian life and had them go to the church and make the repairs. Additionally, the church was expanding its building. Since wood was so expensive, it was cheaper to build walls out of cement blocks. Although they had the cement, they needed sand from the nearby river to mix in, but they had no way to get it up to the church. I contacted our Service and Supply Battalion and sent a 5-ton truck down to the river to haul the sand. There was also a medical clinic in the village that was in need of supplies, so I went to our Battalion Surgeon and convinced him to give any of his excess medical resources to the clinic. Finally, there was a local police station that was no larger than a small one room building; we called it a police box. It was heated by a single diesel stove, and they often ran out of fuel in the wintertime. I had our Motor Pool supply them occasionally with 5-gallon cans of diesel.

My soldiers used to joke that I could have run for Mayor of Chang-Pa-Ri and won. An incident soon occurred that tested their claims. U.S. Army units on the DMZ owned the night because we had a limited number of night vision devices, which we called starlight scopes. In the 60s, this technology was highly classified because it allowed us to detect infiltrators. One afternoon, I received a call from one of my companies. One of our trucks drove through the Vill with a starlight scope in the back of the truck, and the scope disappeared. Korea had "slicky boys," nicknamed as such because they could quickly steal anything, even from a moving vehicle. If this technology fell into the hands of North Koreans or Russians, it would be a disaster. Careers could have ended, and violence could have escalated. I immediately drove to the Village, met with the Police Chief, and informed him that a black box had been stolen at noon that day. I had built a solid relationship with the Police Chief by helping them and the community, so I didn't have to divulge the top-secret contents. I told him I didn't care or need to know who stole it. I just needed it back. Within a few hours, I had the box back. Through the Chang-Pa-Ri fund and relationship-building, the trust and care we built in the community paid off.

Within our company compound, I also ensured that every soldier was appreciated and treated equally with respect. Our U.S. Army units in Korea were assigned Korean Army soldiers, mostly lower-ranking enlisted, who served in our Infantry units. They were called Korean Augmentation to the United States Army, or KATUSAs. KATUSAs were generally well-disciplined and excellent soldiers, unlike the rag-tag Korean soldiers who were called to duty in the Korean War. Our company compound had an NCO club that served beer, soda, and the typical American bar food of hamburgers, fries, and sandwiches that soldiers indulged in. But my KATUSAs had no Korean comfort food, so I brought in a Korean cook to provide Korean food and wine for them. They were able to feast on kimchi, rice, fried dumplings, and all the other fixings that reminded them of home, just like the American troops had. This demonstration of care brought our entire unit closer together. The KATUSAs were members of our unit too and deserved it.

When I departed Korea, I was honored to receive numerous letters of appreciation. The Korean Army recognized my work with the KATUSAs. The local village officials recognized my contributions to the school and hospital. The church also provided thanks. And I even received recognition from the National Police.

But this recognition does not just fall on me. My commanders supported all of my actions and our soldiers did likewise. The smiles of appreciation were enough of a reward, and I hope to have changed some ideas about ugly Americans. Moreso, I was able to see how my leadership principle of Care worked firsthand — both with Americans and Koreans. I was blessed to be in Korea. My first deployment was just the genesis of my lasting connection to the country.

CHAPTER 9
DISILLUSIONMENT: A PAUSE IN ACTION

Shortly before the end of a deployment, we would fill out a form, referred to as a "Dream Sheet," in which we would write our desires for our next assignment. Being a young airborne soldier, I requested assignment to either Fort Bragg or Fort Benning, where paratrooper units and training were located. As an RA officer, it should have been a no-brainer. But instead, I was assigned to the Army Infantry Recruit Training Center at Fort Lewis, Washington, near Tacoma. The pattern of requested assignment denials was continuing.

This rejection was creating a chink in the armor of my decision to make the Army a career. Once I make a decision, I am all-in. But as I was denied assignments, I was growing frustrated by these derailments. I refused to play the victim, but I worried about the prospects of my career.

While in Korea, I had also observed another pattern that was chipping away at my trust of the Army leadership. I noticed that some leaders were more interested in their careers than the unit mission or the welfare of the troops. As someone at the beginning of his profession, it was imperative that I had faith in a system that would generate the type of leadership in which I would have the utmost of confidence. As the Brigade Adjutant in my last assignment, I'd had an insight into a higher level of system and personnel decisions, and the picture was not appealing. In the Army, we have annual individual performance ratings referred to as Officer Efficiency Reports, or OERs. As the adjutant, I reviewed all of the OERs for the Battalion Commanders prior to signature by the Brigade Commander, the rater. I knew the qualifications, accomplishments, and abilities of our four Battalion Commanders. If I were the rater, I would have rated them 1, 2, 3 and 4. The actual ratings turned out to be 4, 3, 2 and 1, totally opposite from mine. The ratings revealed a growing trend in the Army that favored managers, not leaders.

Although I was initially disappointed to be derailed to Fort Lewis, Mr. Mann and his wife, Helen, were delighted to welcome me to Tacoma and offered for me to live with them until I could get settled and find a roommate and apartment. Their kindness and generosity transformed my experience into a positive and pious one. They truly witnessed their faith in their lives, with a loving family and their involvement with the local community and church activities. This was a particularly important time in my life as I had strayed away from fellowship with other believers, violating two of my leadership principles (Example and Balance) in my own life. They ensured I ate my dinner meals with them, always giving grace before each one, and attended services at their church. They even had a special family member, their pet Chihuahua, Juan. Having lived in tenement apartments my entire life, I never had pets. Juan was a very smart dog and a great watch dog. He was able to do tricks and was very friendly to all family and invited guests, but he always warned us and barked whenever anyone approached the house from any direction. He was a true lap dog, and he and I bonded quickly as good buddies. Harold and Helen Mann were instrumental in encouraging and equipping me in my spiritual life.

I had two roommates during my tour at Fort Lewis. The first was 1st Lieutenant Donald Ide. Don was a fellow Infantry officer and was a Washington State native. Don got orders to Vietnam and joined the 25th Infantry (Tropical Lightning) Division. He was killed in action in May, 1969. On June 27, 1969, Life Magazine published the photos of the more than 200 soldiers who were KIA (killed in action) in just one week in Vietnam. The week selected was the week that Don died, and his photo was included. This publication became a rallying cry for the anti-war movement.

Meanwhile, my former roommate from college, Don "Lank" Meyer, had just recovered from wounds sustained in Vietnam at Fitzsimons Army General Hospital in Colorado, and was looking for a new assignment. We had kept in touch, and he got orders to be assigned to the Training Center at Fort Lewis. Lank arrived just after Don Ide had deployed, so he moved in with me. We became roommates again, still single, a little older, but not much wiser. Don was nearly six-foot-four, whereas I was on the shorter side, so we looked like a pair of cartoon character best friends. As we served and lived together, our deep and comforting friendship was one of the closest bonds I've ever had.

When I arrived at Fort Lewis, I was assigned to the Commanding General's staff based on my experience as a brigade staff officer. I was given the position of Secretary of the General Staff, or SGS, authorized as a Major. The General has a staff of Lieutenant Colonels that is led by a Chief of Staff, a Colonel. The function of the SGS was to coordinate the calendars of the General and Chief of Staff and, more importantly, review and coordinate the staff officer actions and recommendations. Many soldiers have never heard of the position, much less know what it does. When done correctly, an SGS can save time for the General and Chief of Staff and help prevent the staff officers from submitting sub-par white papers or recommendations that have not been fully coordinated with the other appropriate staffs involved in the issue. My Pershing Rifles and Korea staff experiences made me a very effective SGS.

But I could only sit behind a desk for so long, so after four months, I requested to be assigned as a company commander for one of the Infantry training companies. Prior to making the request, I had dealt with most of the battalion commanders and knew which ones were best. I then approached one of them when a company command position opened up in his battalion, and he immediately agreed if the Commanding General would approve. The General knew that company command was exactly what I needed at the time, so he approved the transfer. I was moving back to my preferred role, leading and serving alongside other soldiers, but before my tenure as SGS ended, I had the opportunity to meet one man who changed the course of my career, leadership, and life.

Lieutenant Colonel David Hackworth reported into Fort Lewis as a new Battalion Commander. I was the first officer he met at the headquarters due to my SGS job. I knew who Hackworth was, a living Infantry legend. I had already read his informative handbook on small unit tactics and guerrilla warfare, but I tried to conceal my excitement and followed professional regulations. I immediately stood up and said, "Welcome to Fort Lewis, LTC Hackworth, the General will be with you shortly."

Hackworth, as all Veterans will tell you, arrived early to guarantee he would not be late. In practical military terms, combat operations have key times for certain actions to happen by all units in an operation. If one unit fails to do their part at the time stated in the operations order, it could mean the failure of the total operation resulting in loss of life.

I have always followed that policy of never being late — perhaps, I will admit, to an extreme. My children used to complain about always arriving early, but I taught them that it is disrespectful to the other person to be late. Time is something precious that you cannot give back to someone if you are late. Without hesitation, LTC Hackworth responded to me, "What are you doing sitting behind a desk? If you want a company, I'll give it to you." I answered that I was honored, but that I had come to the same conclusion weeks earlier and had already secured a company command from another Battalion Commander. He was not happy but understood I had given my word and had to stick by it. I later learned that Hackworth had an uncanny knack of sizing people up literally on the spot. In his view, you were either a "stud" or a "dud," and there was nothing in-between. Thankfully, he quickly decided that I was a "stud" based on my eager but respectful response. I observed this phenomenon for over 30 years, and LTC Hackworth's intuition was always accurate.

Tough training cycles were necessary to prepare our soldiers for the brutality of Vietnam. As it turned out, my roommate Don Meyer was already assigned to be a company commander in the Battalion that Hackworth took over, so I was debriefed on his first-hand experiences. Hackworth pushed his recruits harder than ever before. He rightfully maintained that we were not doing our recruits any favors by not pushing them to their limits in training. Intense and harsh training would pay benefits and save lives in combat. He was, as always, spot on. I had the same policy, influenced by my samurai teachings. Ninety-nine percent of Fort Lewis graduates were shipping to Vietnam, so no slack could be permitted in our standards.

We were short training officers in my battalion, so I was leading my company with no officers. But I had great non-commissioned officers; my Drill Sergeants and instructors stepped up. Training companies are evaluated by the standard scores their trainees attain at the end of training in physical fitness, marksmanship, land navigation, obstacle course, road march, and first aid. Our company set records for the highest scores ever attained up to that point.

MACV Advisor II Corps. Pleiku, Vietnam. 1969

My acceptance of the other command proved to be beneficial for LTC Hackworth, myself, and one lucky non-commissioned officer. Our training battalions were short field grade officers, so our Battalion Executive Officer position, authorized as a Major, was vacant. One weekend, our Battalion Commander was on leave, so I was the acting Battalion Commander. Our battalion and LTC Hackworth's battalion shared a company street on Fort Lewis. That evening, I received a call from the staff duty officer (SDO) that we had a problem. One of LTC Hackworth's Drill Sergeants had beaten up one of our trainees. I told him not to call the Brigade SDO until I got back to him. I immediately called Hackworth, and we met at my battalion headquarters.

We called in Hackworth's Drill Sergeant, who was a career Infantry soldier, Vietnam Veteran, and airborne. When asked what happened, he explained that he was driving his car down the company street when he saw my company trainee walking with his hands in his pockets, a definite no-no which we referred to as "Air Force gloves." The trainee's immediate response was something about the Drill Sergeant's mother. The Sergeant immediately jumped out of his car and grabbed the trainee by the collar and took him inside a nearby day room and bounced the trainee off the walls a few times to get his attention. He didn't cause any broken bones or bruises other than to the trainee's ego. In the room was just one other person, one of my NCOs, a newly-promoted "shake and bake" sergeant. We were so short on NCOs during this time that the Army fast-tracked soldiers to become Sergeant E-5s only 90 days after completing their advanced Infantry training. They were also called "instant NCOs" and were really learning on the job.

My company trainee was a so-called "jailhouse lawyer," someone who knows enough about the Uniform Code of Military Justice (UCMJ) to be dangerous. He immediately announced, "You can't do that to me. I'll have you court-martialed."

The Drill Sergeant replied, looking at the other NCO in the room, "You don't have a witness. Right, Sergeant?"

My instant NCO responded, "I saw the whole thing." The Drill Sergeant pleaded with his Commander that he was simply applying NCO justice and expected a fellow NCO to have his six (his back). But since the instant NCO had been rushed through his training, he did not yet understand these unspoken Army rules.

Trainee abuse was one of the worst offenses that could take place at an Army Training Center. It could have easily cost that Drill Sergeant his entire career, not to mention time in jail. Hackworth and I conferred on the situation and the severe consequences of our decision, and then we called the trainee in.

Since I was his Commander, I ran the meeting with LTC Hackworth sitting on the side. The trainee's version, as you might suspect, was quite different. He was simply walking down the street minding his own business, and this mad Drill Sergeant took him into the company day room and physically abused him (note the legal jargon) and, to top it off, he had a witness, the NCO from our unit.

"I understand your outrage, and as your commander, I will make sure that you receive justice to your satisfaction," I told my trainee. Yet, this incident was taking place during the end of his training at Fort Lewis, and the Christmas holidays were coming up. All the trainees were looking forward to being with their families before shipping out to Vietnam. I asked him, "What do you want to be done?"

"I want the Drill Sergeant to be court-martialed," he replied.

"I can do my best to make that happen, but these proceedings can take time, and you'll have to be a material witness and stay here for the trial," I informed him. I could immediately see what was going through his mind. If he proceeded, he wouldn't be home for Christmas and might ship directly to Vietnam from here.

LTC Hackworth immediately chimed in: "I might have a

recommendation that would still accomplish punishment under the UCMJ, but it would require your trainee's approval. It would involve Article 15 of the UCMJ," and then proceeded to describe the maximum punishments that he could impose as a Battalion Commander: reduction in rank one grade; up to twelve months in confinement, forfeiture of a month's pay, and an official adverse record in his personnel file.

"We want you to be totally satisfied," I told the trainee, and said I could proceed with the court-martial. LTC Hackworth and I were indirectly trying to convince him otherwise, but we wanted the trainee to believe it was his own idea, in his best interest.

The trainee immediately responded that he trusted LTC Hackworth and preferred the Article 15 proceedings. We asked him again and he repeated his total satisfaction with Article 15. I then told him we would proceed as he wanted and I dismissed him. I held back my urge to then address his actions in the matter, but I didn't want to jeopardize the resolution. We approached the problem indirectly to find a preferable solution, one that would satisfy all parties — the trainee, Drill Sergeant, LTC Hackworth, and myself.

LTC Hackworth then called his Drill Sergeant back in and announced he was busting him one rank, ordering a forfeiture of one month's pay and three months in confinement — all suspended for six months. The suspension meant that if he kept his nose clean, the entire matter would be stricken from the record. Thus, a fine combat NCO's 20-plus year career was saved.

Through this indirect approach to problem-solving, I witnessed why LTC Hackworth's leadership had become legendary. He understood how to adjust tactics to resolve problems, trained subordinates to step up, and above all, cared about the outcomes for his troops. I was lucky that this wasn't the only incident where I'd learned from his leadership first-hand, but it was just the beginning. This resolution cemented the professional connection with LTC Hackworth that would lead to me finally getting to Vietnam.

CHAPTER 10
COMBAT EXPERIENCES: FORGED IN FIRE

My active duty in two combat tours included numerous contacts with the enemy, but there were three special events that merit mentioning.

After serving as a platoon leader in Korea, I was settling into my new position as the Battalion S-1 at the Battalion Headquarters when all hell broke loose. We were notified that the A Company compound had been attacked and sustained numerous casualties, including WIA and KIA (wounded and killed in action). It was very early in the morning before sunrise, still dark outside. I ran to the Battalion Medical Dispensary and hopped on one of the ambulances on its way to A Company. Helicopters were not permitted in the DMZ, so we had no medical evacuation (MEDEVAC) helicopters.

We were up against North Korean forces that were all hand-picked Special Operations Forces. Their mission was to infiltrate through the DMZ into South Korea to create havoc and disrupt lines of communication and supply. They also had a mission to assassinate top government leaders, including the President, who lived in the official presidential residence in Seoul, nicknamed the Blue House. The Special Forces' standard operating procedures utilized quick engagements of targets of opportunity, then directed them to vanish into the populace.

The North Koreans knew exactly what they were doing in this

well-executed sapper attack. Sappers employ explosives. In this case, they breached the outer compound perimeter undetected. Then, they set up satchel charges around on the steel Quonset hut barracks and detonated the explosives with soldiers sleeping inside.

By the time we arrived with the quick reaction force (QRF), the North Korean Special Forces were long gone. Our 18-year-old medics had skillfully triaged the wounded and were doing their best to save the lives of the most seriously wounded. Three had already died. Our Battalion Surgeon and his medics immediately jumped in. I was always amazed by how much responsibility we place on the shoulders of the young medics, affectionately called "Doc" by we Infantrymen. Just by their sheer presence, they are providing life-saving medical and psychological aid, and despite the immense pressure of holding someone's life in their hands, they perform with composure. They are among the most courageous soldiers on the battlefield.

Two and a half years later, I was in Vietnam with LTC Hackworth. While I was commanding a Battle Company in his Battalion, I earned a priceless honor from LTC Hackworth: He gave me my nickname in the military. He couldn't pronounce Mukoyama, so he called me "Mook." I called him "Sir," though he was affectionately referred to as "Hack" by the troops.

Hack had a great knack for making our troops feel special by creating names or mottos only used by our unit. Instead of the usual company designations using the phonetic alphabet of Alpha, Bravo, Charlie and Delta, our companies became Alert, Battle, Claymore and Dagger. When he first took command of the battalion, he changed the name of the Fire Support Base from Dickie to FSB Danger. And most importantly, he named our battalion the Hardcore. We were also a recondo unit, which refers to LRRPs, or

Long-Range Reconnaissance Patrols. Elite units also had military greetings and responses unique to their units when officers entered a unit. This was a tremendous builder of esprit de corps, or unit pride. For the Hardcore, when an enlisted soldier met an officer, they saluted and said, "Hardcore Recondo." The officer would appropriately respond, "No F***ing Slack." I actually saw Hack walk up to a wounded soldier on a stretcher and the soldier would salute and say, "Hardcore Recondo, Sir." These nicknames served as a constant morale boost, and even in the toughest times, everyone was proud to be hardcore.

Amidst the hardships of combat, I was also blessed to have

Proud to be Hardcore with flying helmet. FSB Danger. 1969

one of the greatest spirit boosters — camaraderie with my best friend. The Good Lord provides people in your life that provide a special comfort during the most difficult of times. As mentioned earlier, my former roommate, Don Meyer, was in the ROTC program and was also commissioned as an Infantry officer out of college. Our humble beginnings and shared interests initially bonded us as roommates at the University of Illinois. We both joined a military society called Scabbard & Blade. Don was from a small farming town, and we both were so limited in funds that on Sunday evenings when the dormitory did not serve meals, sometimes we would treat ourselves to a Chinese restaurant in Champaign, where all we could afford was a single order of shrimp with lobster sauce and one beer each. I still remember we used to count the shrimp. One for you and one for me; two for you and two for me. Such experiences bonded us as brothers.

After graduating, Don joined the 101st Airborne in Vietnam. He was wounded and recovered at Fitzsimons General Hospital in Colorado. Once we were reunited at Fort Lewis and became roommates, our friendship was quickly reinvigorated. He then joined Hackworth with the Hardcore and commanded Claymore Company. Then I joined the Hardcore and commanded Battle Company.

Our third reunion in the Mekong Delta was the best of times and the worst of times. For years, the two of us had set our sights on serving our country in on-the-ground combat, but the reality of the Vietnam War was brutal. Though we were in the hands of great leadership under LTC Hackworth, Don and I feared for our futures. The next day was not promised. The one thing we could trust was each other. When we were on combat operations, we both knew that the other would go through hell to come to the aid of the

83

other. I have never felt as close to another person as I did to Don because his comradeship was the biggest comfort in battle. Once we were reunited after an operation, we would have the widest grins plastered on our faces, knowing we had both made it through. God undoubtedly placed Don in my path, and we forged a true brotherhood.

Vietnam was far different from Korea because I was there with one mission — defeat the enemy. As opposed to the Chang-Pa-Ri fund in Korea, I didn't have the time, energy, or resources to support the community. We were not there to win over hearts and minds. We were fighting the enemy in a guerrilla war. LTC Hackworth briefed us on tactics he had learned in previous deployments, but this form of irregular, insurgency warfare was less predictable and organized, more chaotic and treacherous.

As the leader of Battle Company, I was leading operations in these gruesome firefights. In one particularly ruthless battle with the Viet Cong, we had killed many. There were three dead bodies at my feet.

A unit is most vulnerable right after a victory. It is just human nature to let your guard down and breathe a sigh of relief. But as the guy in charge, the pro, I knew that and acted accordingly. I was on my radio with my platoon leaders, barking out orders and telling them to take care of their wounded, reorganize their units, redistribute ammunition, and watch for enemy avenues of approach for a counterattack.

But suddenly, I stopped. I glanced down at the bodies at my feet and realized that something had happened to me, something had hardened my heart. Only moments earlier, these bodies were alive human beings, children of God, with families and loved ones. They were fighting for something equally as important to them as I was fighting for myself. And yet I was treating them like bumps on a log. I then remembered Jesus's Sermon on the Mount in which he told us to love our enemies. So in the midst of all the mayhem, the so-called "fog of battle," I stopped and said a prayer for the three Viet Cong and their families.

But I was also saying a prayer for myself.

Now I didn't have a grand ceremony, and didn't get on my hands and knees. All of this happened in my mind and probably lasted about 30 seconds, but it is something that has remained in my heart and mind ever since. At that moment, God had got my attention for a reason. Though I did not fully understand His lesson yet, this prayer healed me from an invisible wound of war.

About six months later, I became an advisor for the Vietnamese Army serving with the Army of Vietnam (ARVN) Second Corps, known as II Corps. Vietnam was divided into four major Corps areas from north to south, numbered with Roman Numerals: I Corps, II Corps, III Corps, and IV Corps. LTC Hackworth had transferred to II Corps to become the Senior Operations Officer, or G-3, advisor after receiving his eighth Purple Heart. The Army took him out of the field because they did not want the enemy to gain a huge morale boost by killing one of our greatest Infantrymen of all time. A month after Hackworth transferred, President Nixon started the withdrawal of our Infantry units from Vietnam, and the Hardcore was one of the first to be selected. I could have taken my company back to Schofield Barracks in Hawaii. But I was a young stud lifer and had only been in country for five months, so I declined to return. I contacted LTC Hackworth and asked him if he could use me. The next day, I had orders to report to MACV (Military Assistance Command Vietnam) Team 25 in Pleiku where Hackworth was a member. I was assigned as the II Corps G-3 Plans

Advisor, authorized as a Major.

Both my job and war's objectives had changed. The United States was transitioning its active-duty forces to a more supportive role, focused on providing security and winning over hearts and minds. The South Vietnamese would become the main fighting force. With this new, exciting mission, I was responsible for organizing plans that were both militarily and politically feasible. The II Corps Assignment was a similar level to the Pentagon, and the duty was not very dangerous. Occasionally, the North Vietnamese Army (NVA) would rocket our compound to keep us on our toes, but it was a far cry from walking in the rice paddies and jungles of the Delta.

In November 1969, I was assigned to accompany the Corps Deputy Senior Advisor, a full Colonel, and the Senior Enlisted Advisor, SGM Robert Dondero, on an inspection trip of Special Forces Camps on the Border of Laos and Cambodia. I was the young Captain coming along to take notes.

Many of the Special Forces Camps were occupied by Montagnards, indigenous peoples of the Central Highlands and fierce fighters who opposed the North Vietnamese Army. The camps housed the Army of Vietnam forces and their families. In the past, the NVA were known to throw grenades into bunkers when the ARVN forces were away, knowing there were women, children and old people within.

When we landed on the first camp, we immediately came under attack. A rocket was shot, landing just 20 meters away from me. But woefully, it hit the Senior Enlisted Advisor. He was killed. The backstory was even more gut-wrenching: The SGM was scheduled to go on R & R (Rest and Recuperation leave) the following day to Hawaii to meet his waiting wife to celebrate their 25th wedding anniversary.

I, being a Captain, outranked the SGM. I could have ordered him not to get on that helicopter. But I didn't. I carried immense, buried survival guilt for decades.

Combat was gruesome, heart-rendering, guilt-ridden. My deployments in Korea and Vietnam tested my leadership, bravery, and humanity. But I also witnessed men of courage, brothers, and leaders come together for our duty — to protect and serve the finest country in the world. I've shared just three examples of the uncertainty of being in combat. Such moments are unique and never replicated. One cannot control what happened in the past or what will occur in the future, but we can control what our reaction is in the present. As I entered back into the civilian world and into my life's next phases, I always kept these lessons in combat warfare, LTC Hackworth's leadership, and God's divine timing close to my heart.

CHAPTER 11
MOVING ON WITH LIFE: LIFE AFTER
THE ARMY—NOT QUITE

In active duty as a Regular Army officer, I was proactively pursuing advancement through the meritocracy of the military. Yet for the first four years, I never once received an assignment I requested. When I initially entered active duty, I went through my Infantry Officer Basic Course (IOBC) and was an honor graduate. Immediately after, I volunteered for jump (airborne or paratrooper) school and Ranger school. Army Rangers are elite Infantry soldiers trained in patrolling in difficult environments, from swamps to mountains. I wanted to be an airborne ranger. At the time, it was mandatory for West Pointers to take Ranger School, so they took up all of the quotas. But I had a solution. After airborne school, I would apply for Pathfinder School. In Airborne units, the Pathfinders jump prior to the main unit and recon and prepare the landing area. I reasoned that by the time I completed Pathfinder training, the West Pointers would be gone, and I could slide into the next Ranger Class. But my request for Pathfinder was denied too. So, I volunteered for Vietnam and was sent instead to Korea. As I have mentioned, when I completed my tour in Korea, I requested assignment to either Fort Bragg or Fort Benning for assignment to an Airborne unit; instead, I was sent to Fort Lewis, Washington. Each plan I attempted was rejected.

At that time, the Army came up with the Foreign Area Specialty Training Program, or FAST (it later became the Foreign Area of Operations, or FAO, Program). By 1968, the Army realized it had been caught with its pants down in Vietnam. There were practically no officers knowledgeable about the politics, history, or language in that area of the world. In response, they developed the FAST program to produce in-house Army experts in every area of the world. So here I was — a young man with a master's degree with a concentration in Japanese and Chinese history, including Japanese language; ethnically I was Japanese American; I had done a tour in the Republic of Korea and received awards from the Korean and Japanese Military, and to top it off, I was a Regular Army officer. So, I called the FAST program manager at the Pentagon.

"Is there a possibility for me to join the FAST program?" I inquired.

Immediately, the first words out of the Pentagon manager's mouth were, "You haven't been to Vietnam yet."

"You are correct, but if you check my record, you will find that I volunteered for Vietnam. I was sent to Korea instead, which was not a walk in the park," I replied.

"But you have not been to your officer career course either," he responded.

In essence, he was telling me to check with him after a tour in Vietnam — if I made it back — and after a year in the officer advanced course. Basically, he said, "See me in three years." He gave me no encouragement by recognizing my fit for the program. I started to seriously consider leaving the Army when my initial commitment of four years was completed later in the year. But around this time, I received a letter from LTC Hackworth.

"Mook, we have a war going on. If you want a company, it's

yours," LTC Hackworth wrote.

It was a dream come true. To command a company in combat, under the legendary Hackworth, was an opportunity I couldn't pass up. It was like winning the Super Bowl for an Infantry Officer.

Once I was in Vietnam, Hack knew the Army was losing me. He sent a letter to his contacts at the Infantry branch and told them they were in jeopardy of losing an outstanding officer, and they should show him personal interest, so they might be able to save him. That initiative was a reflection of the man Hackworth was. As a leader, he always stepped up to care for his men.

So there I am in the rice paddies, and I get a letter from my Infantry branch manager. I had never been in contact with my branch because I'd naively thought that if I performed well, the Army would take care of me. When I opened the letter, the first line read, "Dear Captain Mukoyama, we have carefully reviewed your personnel file and believe you have excellent potential for graduate school." There was one problem: I already had my master's degree! I showed the letter to Hackworth and told him I wasn't going to put the next 20 years of my career in the hands of such incompetents. Serving with the Hardcore was one of the highest honors in my life, but my mind was made up: I would leave the regular Army after this tour in Vietnam.

Upon returning to the States after Vietnam, I would embrace a new mission in my next phase of life in the "real world" (as we soldiers called it). My immediate plans would include getting reconnected to my church, finding a job, and meeting girls.

I submitted my resignation as a regular Army officer, and my terminal assignment was at Fort Sheridan, Illinois, where I served for my final three months on active duty. When I returned to my home state, one of the first things I did was attend my church in Chicago.

My church, Tri-C, were strong supporters of my active-duty tours, especially in Vietnam. Many of the members were World War II Army Veterans, and they sent me thoughtful packages and encouraging letters when I was overseas. I cherished my church's care packages, and the most valued gift, small packets of soy sauce, was like liquid gold.

But most importantly, I knew that they were praying for me. One should never underestimate the power of prayer, and I never did. I was thousands of miles away from home, facing constant danger and uncertainty. But the knowledge that people were praying for me was the ultimate comfort. I knew I wasn't alone.

The Tri-C Church welcomed me with open arms. The nisei Veterans made me feel right at home. These fellow Infantrymen, who had fought 20 years prior in the Pacific, were eager to hear my stories from combat. My church asked me to teach the teenage Sunday school class. I was delighted to do so as I was still a young guy, not that far removed from that age cohort. This position also gave me another purpose, which helped reintegrate me into American society. I was back in the flock.

But due to the war protests, my church's generosity was not the norm. Regrettably, many churches missed a golden opportunity to support those serving our country, regardless of one's political feelings or the ongoing news cycle. Everyone should be anti-war. In fact, service members and their families are among those who are most against war, because they have to bear the brunt of the wounds and death. Soldiers took an oath to support and defend the United States, so they were only following orders and doing their job.

This new American society was incapable of separating people from policy, or respecting service and sacrifice. The anti-war group mentality had permeated the country's political opinions, especially on the coasts. My first inkling came when we landed in California from Vietnam. We were advised to change out of our uniforms as soon as possible. But being the proud airborne Infantry officer that I was, I disregarded that advice and immediately caught a plane to Chicago wearing my uniform. However, the majority of my comrades received rude awakenings. They were given the one-finger salute, spat upon, and called baby-killers. One of my Catholic friends went to his priest for consolation, only to be told he was going to hell because he fought in Vietnam. My friend immediately turned around and never went back. The California environment was toxic, and my personal anger and resentment was burgeoning.

And it gets worse: Jane Fonda was cast as a protagonist and treated like a hero in San Francisco. The definition of treason is giving aid and comfort to the enemy. The propaganda coup was given to the North Vietnamese. A movie star was wearing a North Vietnamese Army (NVA) helmet and gleefully sitting in an anti-aircraft gun that shot down American war planes. That certainly met the criteria for treason. The service members, me included, hated Fonda so much that urinals had screens with her photo. Since then, my maturing faith has thankfully given me a peace to forgive those anti-war proponents, regardless of their subsequent half-hearted public relations comments. They have never, to my knowledge, included a full, unequivocal apology nor an attempt at some kind of restitution to Veterans. But I have learned that I can never and should not forget — but I can forgive. My God has forgiven me of my sins, so who am I not to forgive? Note, I did not say judge. Only God can render final judgment and know their heart. More importantly, one cannot let such anger and revenge fester, as it only exacerbates the situation. Once I forgave them, I put that hatred out of my heart, and a huge weight was lifted off my shoulders. I will not let the past control my present.

As my fellow Veterans and I witnessed this abhorrent public treatment, we swore to never let this happen again to future servicemen.

I firmly believe that the favorable treatment of returning Veterans today is in good part a result of those efforts. My generation never heard the phrase, "Welcome home." As a result, many Vietnam Veterans did not seek help for their invisible wounds, and instead, just buried them to fester for decades.

A lot of people are familiar with post-traumatic stress disorder and traumatic brain injury, but very few are knowledgeable about this new phenomenon — *Moral Injury*. Research has shown that it is contributing to the high suicide rate.

From the day we are born onward, we are developing a moral code, whether it's from our family, religion, community, or friends. When you join the military, a warrior code is superimposed on your personal code, and in fact, transforms it somewhat. You might have to participate in operations that violate your personal code, such as killing or causing death. You don't have to be the person who pulls the trigger, but you could witness or feel that you could have prevented it. Perhaps you lay witness to another unit killing civilians or are forced to handle body parts. At that time, you sustain invisible wounds — a moral injury. You cannot see these wounds. In the military, you are constantly moving too, from Point A to B to C, so you don't have time to stop and reflect. You bury it. It becomes unresolved grief and shame. Especially when you return to the States and you come back to a community that can't understand what you've gone through, this pain will boil up. Anger, guilt, depression, suicide. From Shakespeare's *Hamlet* to Homer's *Iliad*, wars have forever changed the brave men who have fought. These deep-rooted moral injuries are now resulting in suicides in older age, when one has time to reflect on life events. The vast majority of Veteran suicides occur in those over fifty years old.

This tragedy was driven home in an experience that my wife and I shared in 2018. I was giving a presentation on Moral Injury at a graduation dinner at a medical school in Lebanon, Oregon, one of the most patriotic communities in the country. When we arrived at the hotel, there was a message on the marquee saying, "Welcome, General Mukoyama." When we were in the elevator going up to our room, there was a hotel employee who noticed the Vietnam Veteran cap that I wear every day. I make sure to wear that cap every day to let people know that we have Veterans everywhere in our community.

"Hello sir, I am a Vietnam Veteran, too!" the hotel employee immediately pronounced. You know what happened next. Whenever two Veterans meet, we immediately start talking. I discovered that he was in the Army — obviously a man of great character — and was in Vietnam the same year as I was there. We continued the conversation in the hallway, and he said, "Hey, did you hear that there's going to be a General here?"

To which I shyly nodded my head and said, "Ah, yeah. That's me." I gave him my card, shook his hand, and said, "Welcome home!"

I gave my speech that evening, and the next morning, the hotel employee came running out as I was packing up my luggage into the car. He shook my hand with tears in his eyes.

"Ever since I returned from Vietnam, no one has ever said what you said to me last night," he told me. After four decades, no one had ever welcomed him home? I can only imagine the emotional baggage he had carried for over four decades. None of our Veterans should have had to struggle in silence, in fear of the society they served and sacrificed for. If we had accepted our Vietnam Veterans with open arms, ready to heal their internal wounds, we would have saved lives.

When I came home from Vietnam, I was incredibly grateful to be welcomed home by my loving parents. My parents had purchased a two-flat building with my older brother and his wife, and I lived with them upon my return. My parents lived on the first floor, and my brother on the second floor. While I was away, our family gained a new member — another "God Thing." His name was Robert Adams. He and my mother worked for the same company, Lanier Voicewriter, that specialized in voice recording equipment. Bob, who was an Army Veteran and served in the Signal Corps, utilized his communication skills and worked as a service repairman, while my mother was the service coordinator who dispatched all the servicemen.

My mother acted as the mother hen of the service department, and Bob was her favorite serviceman, so they developed a close bond. He had dinner at our house and met my father, and Bob enjoyed hearing my father's stories. He soon became a fixture with my parents, spending evenings enjoying the meals and their mentoring. My parents did not have a car, so Bob drove them whenever they needed a ride. When they bought their first home, the two-flat in Chicago, he helped them move. I was away at graduate school, so I missed out on all the hard work of moving. I then shipped off to Korea and was gone for the next four years. Since we had never owned a home, neither my father nor my brother was handy at fixing basic home fixtures. But Bob was extremely skilled in those areas and did most of the painting, electrical, and plumbing repairs. Most importantly, Bob filled the void that I left as the second son. My brother was there, but he was married and had to concentrate on his family and career responsibilities. Having been in the Army, Bob was able to explain what I was experiencing, especially when I was going through my training at Fort Benning where he had once been stationed and later during my overseas tours. I will be forever grateful for the Good Lord bringing Bob into the Mukoyama Clan.

My next task was to get a job. I applied for a job at the Chicago Police Department that was in the process of re-organizing its force to meet the changing situation on the streets. This was a perfect fit for my recent military experience at II Corps Headquarters, where we were restructuring the Vietnamese Army to take over from the U.S. Forces while simultaneously trying to win a war with the NVA and VC. Also, the police department was as close as I could get to the military job-wise, with its strict organization and code of service to the community. The transition would have been very smooth. Finally, I grew up in the city and knew the neighborhoods because I was an equal opportunity guy and dated girls from different areas. I even had my master's degree to qualify me further. I had an interview but was not accepted.

However, I found my new mission when I joined the U.S. Army Reserves. When I resigned from my Regular Army Commission, I applied for a Reserve Commission, because I had made a personal commitment to serve my country for twenty years. The U.S. Army consists of three main components: the Active Army; the U.S. Army Reserves; and the State Army National Guard when called to active duty. The Reserve components — the U.S. Army Reserves and the State Army National Guard — did not have a lot of combat-experienced soldiers, so I felt that my experiences could be a helpful perspective.

This decision was crucial for me because I was able to maintain my military culture connections. When you leave the military and rejoin the civilian culture, the change in ethics and values makes you feel like you're moving to another world. The Armed Forces have universal core principles of honor, integrity, loyalty and selfless service that is often lacking in the corporate world environment. In the military, you always have a clearly defined mission. The Army Reserves became my new mission, and so the military beat goes on.

CHAPTER 12
HEROES: STANDING ON THEIR SHOULDERS

"Never meet your heroes" is a common warning. But I couldn't disagree more, especially after meeting American of Japanese Ancestry (AJA) Veterans who fought in World War II. My life story has been shaped by their impact and sacrifices, and their mentorship has provided me with the strongest foundation to look up to. I am indebted to them, as is the United States of America.

Although most AJAs were being rounded up and forced into concentration camps during World War II, many men volunteered to serve and prove their loyalty to this country. The military had initially denied them enlistment into the Army and classified them as enemy aliens, but the United States had exhausted much of their forces by 1943 and needed to call upon additional forces. Roughly 10,000 Americans of Japanese Ancestry joined together to form the 442nd Regimental Combat Team. This segregated unit was made up of roughly two-thirds Hawaiian-born nisei and one-third mainland nisei. They first fought in Africa, and their performance was so spectacular that when the other two battalions completed training and joined them in Europe to complete the 442nd Regimental Combat Team (RCT), the 100th Battalion was given the distinct honor of retaining their 100th Battalion designation instead of being renamed the 1st Battalion of the 442nd. These two combined forces were critical in the European Theatre, as they pushed the Germans out of Anzio, Italy, the area north of Rome, and parts of Eastern France. The 100th Battalion and 442nd RCT became the most highly decorated unit for valor for its size and time of service in the history of the United States Army. The unit received over 4,000 Purple Hearts, 4,000 Bronze Stars, 560 Silver Star Medals, 21 Medals of Honor, and seven Presidential Unit Citations.

Others served in the equally valorous Military Intelligence Service, or MIS. This unit of Japanese language linguists served in the Pacific Theater of Operations (PTO) as front-line units, prisoner of war (POW) interrogators, and document translators. Major General Charles A. Willoughby, General MacArthur's Chief of Intelligence, commented after the War that their contributions in breaking the Japanese Imperial Code shortened the War by two years, thereby saving countless lives on both sides.

My association with the Nisei Veterans was first established through familial ties. Richard "Gus" Hikawa, my second cousin's husband, was a member of the American Legion Chicago Nisei Post 1183 and was a past Post Commander. Gus had served in the Military Intelligence Service (MIS) during World War II and served in the Pacific, including the battle of Iwo Jima. Upon returning to the States, he got his law degree and became an attorney at the Traffic Court for the City of Chicago. He joined the Illinois National Guard and eventually retired as a Lieutenant Colonel. My brother and I, when we were children, went to visit Gus at the National Guard Armory on Chicago Avenue and climbed on the tanks.

My entire youth, I was walking among these quiet giants. These were true heroes, but I did not realize nor appreciate it at the time. Despite the prejudices and injustices they experienced from the treatment of many so-called "Americans" during World War II, they chose not to let the past control them. Instead, they were grateful for the blessings we as Americans enjoy every day, and concentrated on marrying, creating families and teaching their children values and work ethic. All of this while contributing to the community through serving.

It was only later, as I learned the history of our country, that I fully appreciated that the opportunities open to all minorities today were a result of their efforts. When Japanese Americans resettled after the concentration camps, the Nisei Veterans' Service fostered the population's acceptance of them. Through their sacrifices in World War II, they set a remarkable example of Japanese Americans' bravery and patriotism. But it is a trait of the Nisei not to brag. It is my mission as a beneficiary — and as an American — to ensure that their efforts are appreciated and not forgotten. I have truly stood on the shoulders of these heroes.

The Nisei Veterans have also fostered my personal development into a leader and future soldier. When I was a high school junior, I was sponsored by the 1183 Nisei Post to attend the annual American Legion Boys State convention in the summer of 1960. The American Legion is a national service organization whose mission is to promote the fundamental values of our nation, as demonstrated in the American Legion preamble that begins, "For God and Country…" The Boys State program encouraged active citizenship through understanding the principles and basic structure of our government. After a competitive application process, high school juniors from all over the state of Illinois met in the summer for a one-week program, at the State Fairgrounds in Springfield, Illinois, and set up a mini-state government with cities, counties, and state elections with two fictitious political parties. We lived in dorms, with each floor representing a city and each building representing a county. Run by Veterans, we had a daily schedule with flag-raising and lowering ceremonies. We had daily inspections of our open-bay dorms. For the first time, I learned how to make the bedsheets and blanket so tight that the inspector could bounce a quarter off the bed. This skill, and many others I learned there, held me in good stead in my later Army basic training.

My attendance at Boys State was a tremendous boost to my patriotism and pride of country. Illinois was the first state in the nation to have a Boys State, and therefore, it is referred to as the Premier Boys State. Political party candidates had to debate the opposing party candidate, and the parties had to develop party platforms. I was honored to be elected City Clerk of my city. There were also additional activities for county sports teams and the Boys State Marching Band, which I joined and played clarinet. This was a tremendous experience for me in terms of leadership, teamwork, and citizenship. And the American Legion subsequently developed a corresponding Girls State Program. Like my experience at the Boy Scout National Jamboree four years earlier, I was exposed to boys from all over the state and from all races, ethnicities, and socio-economic backgrounds. Again, the spirit of Americanism was being realized.

As a National Veterans Service Organization (VSO), the American Legion promotes other patriotic activities in accordance with its charter. Just to name a few, it supports the famous American Legion Baseball program, patriotic essay writing competitions, and community service programs, such as supporting our VA hospitals, outreach programs to Veterans, schools, food pantries, and homeless shelters. The Posts have full flexibility depending on the specific needs in their local communities. And there are numerous other nationally chartered VSOs that are providing essential support to our communities, such as the Veterans of Foreign Wars, the Vietnam Veterans of America, the Military Order of the Purple Heart, and the Disabled American Veterans. I am a lifetime member of all of these organizations and can proudly vouch for them. Even more VSOs have developed from the proud generations of post-Vietnam service members, and are equally worthy of support.

When I returned to Chicago from Vietnam, I was immediately contacted by the Chicago Nisei American Legion Post 1183 with a huge "Welcome home," not afforded to most of my fellow Vietnam Veterans. Initially, the membership of the Nisei Post 1183 was virtually all World War II Veterans, the majority of whom had served in the famous 100th Battalion/442nd Regimental Combat Team (RCT). I got to meet and fraternize with these legendary veterans in the Nisei Post 1183. Although they were physically small, their moral character was considerable. They demonstrated all the great traits of a Nisei, as their humility, loyalty, and perseverance were undeniable.

In the years after Vietnam, the Nisei Post members continued to monitor my military career and encourage me. In 1991, I was invited by Chicago Mayor Richard M. Daley to give remarks at the 50th Anniversary of Pearl Harbor at the Daley Center in downtown Chicago. As an American of Japanese Ancestry, my role as the keynote speaker was very significant because it recognized the historical importance of the event and our nation's unified response.

The following year, the American Legion National Convention was held in Chicago, and the Nisei Post was a proud sponsoring organization. I was honored to make a presentation to the convention attendees and update them on military matters. But equally important to me, I was able to express my gratitude to them for the Legion's influence in my life through their programs and the support of the Nisei Post.

Robert Hashimoto, Mrs. Irene Inouye, Senator Daniel Inouye, Sam Yoshinari,
and I at the American Legion Chicago Nisei Post 1183

After my military career came to an end, I continued to support the Post by being a member of our Post Scholarship Committee that awarded scholarships to high school and college students, based on demonstrated academic and community service accomplishments and an essay based on the service records of the Japanese American Nisei soldiers during World War II. The encouraging aspect of that experience was to see the quality of the young people in our community, many of them being family members of Nisei Veterans, and the pride they received from learning about the sacrifices of these Veterans was both enlightening and emotional for them.

So many Post members have found plentiful success in their careers, while continuing to contribute to our community and nation. One of them was Allen Meyer, a Jewish attorney in Chicago. You might ask how a Jewish Veteran entered into this portion of the story. Allen was interested in Japanese culture, learned to become fluent in Japanese, and joined the MIS to serve alongside the Nisei

103

soldiers. After the War, living in Chicago, he became highly involved in the activities of the Post, serving for many years as the Post Judge Advocate, or legal officer.

Another MISer and Nisei Post member was Arthur Morimitsu. Art served in the China-Burma-India (CBI) Theater of Operations and was very active in national Japanese American Veteran organizations. He got me deeply involved in a national foundation to build a memorial in Washington D.C. to recognize the loyalty and patriotism of Japanese Americans during World War II.

In 1993, Art called me and asked if I was going to attend the 50th anniversary celebration of the 100th/442nd and MIS in Honolulu. Art was in his 90s and asked if I could accompany him. He was scheduled for an intense operation after the event. He was concerned about the surgery, and he felt the celebration might be his last opportunity to be with his Nisei brothers-in-arms. I immediately agreed and went to the celebration event with him. It was an honor to be there not only with Art, but with all of these heroes. Unfortunately, Art's premonition proved to come to pass, with him passing away from complications of his operation shortly after returning. But our trip was the greatest commemoration I could have asked for. I consider our time together in Hawaii a highlight in my life.

As with Art and Allen Meyer, I have been honored to befriend and hear the extraordinary memories of these American Nisei heroes. When I was in Hawaii in 1993, I met so many remarkable Veterans that lived up to their stories of valor. Granted, I had to drag these stories out of them on account of their humility. Being a Major General at the time, I was invited to be on the reviewing stand with Senator Daniel Inouye, an original 100th Battalion Veteran who was awarded the Congressional Medal of Honor. Not only was Senator

Inouye a valiant World War II veteran, but he was Hawaii's first representative in Congress in 1959 and served as Senator from 1963 until his death in 2012. I had met him earlier in Camp Shelby, Mississippi, where we both spoke at the dedication of a monument to the 100th Battalion, 442nd Regimental Combat Team. The Nisei Veterans had trained there prior to going overseas.

In Hawaii for the 100th Battalion & 442 Regimental Combat Team 50th Anniversary. 1993

In the process of bonding with these men, I've had the privilege of hearing their valorous stories and seeing their heroic character first-hand. For example, I was lucky to witness the special relationship between Senator Inouye and another Chicago Nisei Post member, Sam Yoshinari, who was a Lieutenant in the Cannon Company. In April 1945, the two were serving in the European Theatre. Inouye was hit in his right elbow by a German grenade, but Sam was there and immediately applied a tourniquet. Although Inouye lost his arm, Sam saved his life. The War ended three weeks later, but the two men's bond has lasted a lifetime. Whenever I attended a 442nd event that Senator Inouye also attended, he would always make a beeline for Sam to spend time with him. Sam had received a commendable battlefield commission, a Silver Star, for his actions, and he was just as strong of a family man and supported his handicapped wife for many years. She could only travel in a wheelchair, so Sam purchased a van that was wheelchair-accessible, and he would always drive her to his numerous events nationwide in support of the Nisei Veterans. He was his wife's personal caregiver, taking care of her hair, make-up, and clothing for all occasions. She was a beautiful lady, and her appearance was always immaculate. Sam was a true hero in more ways than one. He and I became close friends, and he was one of my mentors and inspirations.

Though there are countless outstanding anecdotes from these World War II Veterans, I am compelled to retell one about Camp Shelby which Senator Inouye personally shared with me. Prior to the War, there was an intense rivalry between the Japanese Americans from Hawaii and those from the mainland of the United States. The Hawaiians referred to themselves as "local boys," but the mainland Japanese nicknamed them "Buddha heads" or "*Buta* (pig) heads." On the other hand, the mainlanders were referred to as *Katonks*, which alludes to the hollow sound heard when coconuts fell and hit their heads. Both sides looked down on each other. The islanders especially resented the mainlanders for making fun of their so-called pidgin English, which was a combination of Hawaiian and English.

They assumed the mainlanders considered them less qualified due to their perceived deficiency in speaking the Queen's English. Many of the Islanders had come from the plantations, where disagreements were often settled with their fists instead of words. As a result, fights were frequent, to the point that the Army was considering breaking up the Regiment.

But one weekend saved and changed the relationship between these Nisei soldiers. Both the mainland and Hawaiian troops were scheduled to have a weekend pass to go to the USO at the nearby so-called "Relocation Center" in Jerome, Arkansas. The Hawaiian guys were looking forward to meeting Japanese American girls, and on the bus ride to the camp, they played ukuleles and sang songs. The mainland soldiers were reserved and probably contemplative. They had not shared their stories about the concentration camps to the local boys, so islanders had no idea what they were about to encounter. Considering the cultural standards of Japanese Americans, the mainlanders may have been ashamed. The Nisei may have also just been determined to persevere and endure the hardship with patience and dignity — *gaman* as we like to call it. Their introversion might have also contributed to an impression of aloofness. But when they arrived at the camp, perceptions changed. Mainland Americans of Japanese Ancestry were being confined to these camps, surrounded by barbed wire with machine gun towers with weapons facing in. Hawaiian AJAs were not universally rounded up and put in concentration camps on the islands.

The Hawaiian guys now understood the realities facing mainland Japanese Americans, and a feeling of respect for the Katonks was born. They asked themselves, would I have volunteered to die for my country when I and my family have just lost our homes, farms, businesses, and freedom, and are languishing behind barbed wire? I too have asked myself this same question and still struggle with an answer. On the bus returning to Camp Shelby, you could have heard a pin drop. All of these soldiers quietly understood that not only were they fighting to help their country win, but also to demonstrate their allegiance and reclaim their freedoms. After that incident, the Regiment came together and became true American brothers-in-arms.

With actor Pat Morita (R). Honolulu, HI. 1993

The Nisei Veterans found camaraderie with their fellow Americans of Japanese Ancestry, but they were also welcoming and inclusive with all of their American comrades. A dear fellow Vietnam-Era Veteran, Carmine Iosue, has a special story in this regard. His father was an Italian immigrant, who volunteered to fight in World War II, and his unit was fighting alongside the 100th/442nd when he was seriously wounded and sent back to a field hospital. His father was heavily bandaged, with a slight build, dark hair, and the name "Iosue," so the hospital assumed he was Japanese American and placed him with the wounded from the 442nd. During the lengthy recovery time, the error of nationality was discovered. But the wounded soldiers recovering together formed a special bond. Carmine's father declined to be separated, and remained with the Nisei soldiers. He later told Carmine that the men of the 442nd would not leave that area unless they were dead, medically evacuated, or the War was ended. After the war, Mr. Iosue maintained close contact with the 442nd Veterans, and he was accepted as one of them. In 1975, Mr. Iosue was invited by the Nisei Veterans in Hawaii to attend the American Legion National Convention hosted in Hawaii, all expenses paid. Unbeknownst to Mr. Iosue, they hosted him and his wife for a dinner in his honor. Years later when I met Carmine, my connections with and knowledge about the Nisei soldiers and our shared Army experiences bonded us forever. In 2021, Carmine visited the islands with his wife, Tina, and I introduced him to the Sons and Daughters of the 442nd, who were excited to meet them. Carmine and Tina are strong people of faith and very generous in supporting non-profit organizations, including Military Outreach USA.

Like many Veteran organizations originally founded by World War II Veterans, the Chicago Nisei Post has reduced dramatically in membership since its days of deep activity, including its highly competitive, and nationally ranked, Drum and Bugle Corps known as the "Nisei Ambassadors." The current commander is a retired USMC Colonel, Bob Hashimoto, who had a combined Active and extensive Reserve 40-plus year career, starting out as an enlisted man. Bob is dedicated to maintaining the honor and tradition of the Post and to making sure that it will be viable as long as possible. He has recruited new members. In 2021, I was elected as the Post Chaplain, fulfilling one of my life-long ambitions to perform spiritual duties in an official capacity. However, I realized years ago that we are all ministers of our faith regardless of academic or seminary qualifications.

After I became a General Officer and traveled throughout the nation giving speeches and presentations to Veterans groups in my subsequent nonprofit activities, I was honored to speak to countless exceptional Veterans. But I must admit that my encounters with Nisei Veterans were the most memorable and rewarding. It was a mutual love fest between us because we were tremendously proud and admired one another. The Veterans wanted to get their pictures with "the General," and I wanted to get my pictures with these living legends. It is not often in life that one can rub shoulders with their personal heroes and thank them for letting one stand on their shoulders. I thank the Good Lord for providing me with such opportunities.

CHAPTER 13
BUSHIDO: EARNING MY AMERICAN KATANA

The importance of the Samurai tradition in Japanese culture cannot be over-emphasized. In the 12th century, the standards and expectations of the warrior class were founded under the *Bushido* (*bushi* = warrior; *do* = way or path), or "Way of the Warrior." It provided a code of honor, to include how Samurais were to conduct themselves in war, peace, and in death (*hara-kiri*). It was strictly for the warrior class. Similar to the Chivalric Code of the British Knights, the *Bushido* code established a level of respect for warriors of all cultures.

And who were these warriors? The emperor reigned at the top of the government. All authority flowed from him since he was considered a divinity. But the actual control of the government had been taken from the court system and put into the warrior class. In 1192, the emperor appointed Minamoto Yoritomo as the first shogun (*sho* = commander; *gun* = troops), the top military commander. The shogun was considered the military protector of the emperor. The rest of the shogunate class, in descending order, was the daimyo, or lord of an area, and the samurai, or warriors who served the daimyo.

Only those in the warrior class were permitted to carry swords, called *katana*. The *katana* were the envy of the world because of their incredible strength and razor-blade sharpness. The manufacturing process was a secret among the sword-making clans, and the actual craftsmen were considered living national treasures.

The power of the samurai over the populace was virtually all-encompassing. Since they were the only ones with access to weapons, the people were at their mercy. Thankfully our U.S. founders created the Second Amendment within the Bill of Rights, but the Japanese samurais had no checks or balances on their power. Their only control was the guidelines of *Bushido*. In fact, there was a saying in Japanese, *kitte suteru*, which translates to "cut and throw away," or "kill and walk away." In essence, the samurai could be the judge, jury, and executioner in one. One could kill a person with his katana and simply walk away.

The shogun held all the actual power in the country. That is until 1853, when Commodore Matthew Perry forced the Japanese to open up their country to trade relations with the United States, thereby ending their self-imposed isolation from the rest of the world. Perry had quite literally forced himself into Edo Harbor — what would become modern day Tokyo —with United States' cannons and firearms. The samurais' swords and archery were no match for the Naval force backed by gunpowder, and the shogun was unable to stop him. The shogun's omnipotence was disproven, effectively ending his reign of power and eventually resulting in the restoration of the imperial authority.

The samurai class's demise was also due to the Imperial Sword Abolishment Edict of 1876. This decree prohibited the wearing of swords in public. The removal of the katana symbol drastically reduced the distinction between the samurai from other common citizens.

After World War II, the Japanese military transitioned to become a primarily defensive force. The former almighty Japanese military was drained of its offensive ability, and thus, most of its power. But the samurai traditions remain entrenched in Japanese culture. The *Bushido* code may have existed for warriors, but this way of life also guided men towards honorable values. And as men like my father migrated around the world, they brought these customs with them. Little did I know that my father would also live to perceive these samurai qualities in his son.

Even though he was thousands of miles away, my father stayed connected with his country of origin. During the 1950s and 1960s, the Japanese Consulate in Chicago was not staffed enough to provide support to visiting dignitaries and business executives from Japan, which was still getting its feet back on the ground after the devastation of World War II. My father, being an independent small businessman, had the time and flexibility to assist these visitors. He would help coordinate their hotel accommodations, pick them up at the airport, and drive them to their appointments. He would often act as a translator for both sides. To my knowledge, he did not charge for these services, because he was trying to encourage an economic relationship between the U.S. and Japan. Over the years, he developed a reputation as dependable and knowledgeable of the Chicago business and government landscape. My father established meaningful bonds with numerous business and government executives in Japan. Even within the Japanese American community in Chicago, he served as President of the Yamanashi Kenjin-kai, which was the club for immigrants from my father's home prefecture of Yamanashi (*ken* = prefecture; *jin* = people; *kai* = club).

During my first overseas duty in Korea, I chose to visit Japan for my two-week R & R leave. Some guys would go to Australia or Thailand, but there was no question where I would be going for my rest and recuperation. I was excited to learn about my ancestors and see the country that my father had left decades earlier. Respect for elders is extremely valued in Japanese culture, and I had an obligation to visit the cemetery where my grandparents were buried in my father's hometown of Kofu. Fortunately, my father's brother, Teruo, had returned to Japan after the war so I could stay with him in Tokyo. Although my father had not returned to Japan since he left in 1918, he notified all his contacts in Japan who he had helped over the decades and asked them to take care of me when I visited. The Japanese phrase is *onegai-shimasu*, or "please," when asking a favor. In these moments, the cultural value of *on*, or debt, clearly manifested.

As an AJA servicemember, I felt as though I was reconceiving the perception of my fellow Americans. My uncle Teruo had warned me not to wear my uniform when I visited Japan. The Vietnam War was heating up, and the U.S.-Japan mutual defense treaty was up for renewal. There was anti-American sentiment via student demonstrations. Frankly, the last thing a soldier wants to do is wear his uniform when on leave, but I was a proud Infantry officer in the United States Army. I ignored the advice. When people saw me walking through the streets of Tokyo, people were doing double-takes. Since America had troops stationed there in the post-War occupation, the Japanese knew what my uniform represented and recognized that I was an officer. But my face and "Mukoyama" nametag didn't compute in their minds. I could visibly see their confusion, as they had never seen anyone who looked like me in a position like mine. But for me, this was a demonstration of democracy in action.

Kofu exceeded my expectations. My uncle and I took the train there, and when we arrived, the Yamanashi Governor sent a limousine to pick us up at the train station. They immediately drove us to the Kofu City Hall, where I was met by the mayor. He personally gave me the keys to the city. The next day, I was proudly featured in an article in the newspaper. I was photographed in uniform, and the title read, *American Army Officer, Son of Local Boy Visits Kofu.* That evening, we stayed at the Prefecture guest house for visiting dignitaries. It was on a beautiful lake at the foot of Mount Fuji, and opened up just for us. I had no idea that all these gracious deeds were going to happen until I arrived. My father had coordinated all of this.

Uncle Teruo's brother-in-law, Negishi Shintaro, had a lot of clout in his own right, as he was a graduate of one of Japan's most prestigious universities and the President of the Boy Scouts of Japan. One day, he visited us at my uncle's home and asked if I would like to meet the Chief of Staff of the Japanese Ground Self-Defense Force, or GSDF. (They cannot call it an Army since their Constitution — that we had imposed upon them after World War II — outlawed war as a means of settling international disputes.) I didn't think he was serious, so I casually replied, "Sure." I was quickly proven otherwise. He immediately picked up the phone, made a call, spoke for a few minutes, hung up and said, "Okay, it's all set. Tomorrow morning you will be picked up here, driven to the Japanese Airborne Division training base, have lunch there, and then be driven back to Tokyo to the Ground Self-Defense Force Headquarters (their equivalent of the Pentagon) and meet General Yoshie, the Chief of Staff, for 15 minutes." Now the pucker factor comes in. I asked myself — what have I gotten myself into? I was just a First Lieutenant, what was I doing visiting with a Four-Star General?

Sure enough the next morning, we were picked up by a military limo and driven to Camp Narashino. I was met by the Deputy Division Commander who apologized that the Commander was not able to meet me. I was taken out into the field where the Airborne Division had a demonstration of paratroopers jumping out of a plane just for me. I then got a tour of their museum that detailed a proud record of wartime service. World War II was the only war that they had lost. There was a world map on the wall with parachute badges from allied nations around the world. I noticed that they did not have a badge from the Korean Army, which I recognized was due to the Japanese Korean animosity. I filed that observation in the back of my mind. During our lunch in the officers' dining hall, I was placed in the Commander's seat.

After I was driven to Tokyo, I had the privilege of touring their headquarters and meeting their Chief of Staff. The moment I walked into the General's office, I noticed that he was wearing a United States Army parachute badge. So being an Airborne soldier myself, we started talking about paratrooper training and jumps. The 15-minute meeting turned into a half-hour discussion. Then some photographers entered the room and snapped some photos of the two of us. The General pronounced, "Lieutenant Mukoyama, I have seen the newspaper accounts of your visit to Kofu. Your presence has had a positive effect on the U.S.-Japanese relations, and I want to demonstrate my gratitude. You are a qualified military parachutist and, as the Chief of Staff, I am authorized to award the Japanese Parachutist Badge to allied soldiers and am therefore awarding it to you."

I was truly humbled and I gratefully accepted. The *Pacific Stars and Stripes* newspaper printed an article the next day with the whole story about an American Officer receiving an award from the Japanese Army. Again, this trip resulted in good press for U.S.-Japan relations during this tumultuous time.

When I returned to my unit in Korea, I immediately went to

our Korean Liaison Officer and told him about my trip. I informed him that the Japanese Airborne Museum had nearly every country's Badge, except for Korea, and asked if I could get a Korean Army Airborne Badge to send to them. Because I had built such a solid relationship with the Korean Army, he agreed without hesitation. I promptly sent it to General Yoshie as a token of gratitude to the Airborne Division.

Two years later, I was visiting Japan again on R & R, this time from Vietnam. Mr. Negishi came again and asked if I would like to meet the Minister of Defense, Nakasone Yasuhiro. This time, I knew he was serious and came prepared. Sure enough, I met Mr. Nakasone the following day in his office. He had briefly attended an Ivy League school in the states and spoke English well, although he didn't speak English to the press. As the Minister of Defense, he was very interested in my activities in combat, especially in Vietnam. Years later, he became the Prime Minister of Japan. When I was promoted to Brigadier General, he sent me a personally written and autographed poem in a Japanese art form of Chinese calligraphy. Called a *shikishi*, this scroll is sent for auspicious occasions. I still have it prominently displayed in our home today.

Commemorative Caligraphy from the Prime Minister of Japan. 1989

My final military trip to Japan occurred in 1992, when I was the senior-ranking officer of a delegation from my headquarters, Training & Doctrine Command, Norfolk, Virginia. By this time, I was a two-star

General, and I arranged to make three special appearances during our one-week conference with the Japanese Military. This annual conference was designed to exchange plans and operations between our two militaries. My appearances were intentional to promote U.S.-Japan relationships, as I had two decades earlier. But this time, I was more deliberate and in a higher official capacity. Additionally, this era was marked by uncertainty, because we were unsure how defense relationships would be affected if candidate Clinton won the presidency. As an AJA, I was in a special position to address the Japanese, so I prepared speeches in Japanese. Although I am not fluent in Japanese, we had a Colonel from the Japanese Army assigned to us at our headquarters in Norfolk. I wrote my remarks in English, then the Colonel translated it into Japanese and made an oral recording, so I could get the proper intonation and phrasing. Then I practiced, practiced, and practiced.

Although the speech slightly differed for each audience, the message stayed the same. I shared that my parents were from Japan, even mentioning their hometowns to get the attention of any soldiers from those areas. Then I talked about my connection with the Ground Self-Defense Force twenty-years earlier. I assured them that the United States was a staunch ally of Japan, both economically and militarily. As two strong democratic countries, we would remain allied regardless of who was the next President.

I gave this speech three times. First, I presented it to the Airborne Division soldiers at Camp Narashino. Just like when I visited the museum twenty years earlier, I received another reception and tour, although this time I noted that the Korean Airborne badge was proudly displayed. Second, I spoke to the students at the Japan Defense University and had a specific meeting with eight cadets. And third, I delivered my speech to the senior officers at the National

Institute for Defense Studies, the future generals and admirals. Finally, I met with General Nishimoto Tetsuya, the GSDF Chief of Staff who later became the Chairman of the Staff Council of the Defense Agency. I had known General Nishimoto for six years because I befriended him when I visited Japan with a touring group of General Officers. He was a two-star General at the time, and we kept in touch over the years.

Commander & Staff 1st Airborne Brigade. Camp Narashino. 1992

My military experiences in Japan were truly unforgettable. Not only was I enlightened on my ancestors' culture and the country's samurai traditions, but my appearances helped to bind the two countries' alliance closer together. A soldier did not have to look a certain way to fulfill his military duty to their country. I was living proof that you could be Japanese in ancestry yet American in citizenship, and still proudly serve your country.

When my tour as commander of the 85th Division Cavalry unit came to an end, the officers and enlisted personnel presented me with an officer's cavalry saber. What was especially memorable about the presentation was the person who gave it to me. It was presented by my Executive Officer, Major Richard "Rick" Rescorla.

For those unfamiliar with Rick, who retired as a Colonel years later, he was a hero both in combat and fighting global terrorism. If you have seen the book by LTG Hal Moore entitled, *We Were Soldiers Once...And Young*, Rick's photo is on the cover. Rick was a platoon leader in the Battle of Ia Drang and his heroism was highlighted in the Mel Gibson film, "We Were Soldiers." But his fame is perhaps equally, if not more, notable for his actions on 9/11 when he was in charge of security for Morgan Stanley and personally led the evacuation of the nearly 3,000 employees from Tower Two and went back in to rescue others when the building collapsed. Morgan Stanley lost a total of only 13 employees due to his leadership and courage. Ironically, his men in Vietnam had nicknamed him "Hardcore," so I was connected to that phrase twice in my life. Rick was a shining example of a citizen-soldier. When I received my cavalry saber from Rick Rescorla, I had earned my American katana. I am so grateful that my parents, especially my father, saw me maintain *Bushido* throughout my military career. When I became a commissioned officer, my father saw a samurai. When I became a General Officer, he saw a daimyo. And when I became the Deputy Commanding General at one of the highest Army Commands in our country, he saw a potential shogun.

CHAPTER 14
WHAT IS AN ANGEL? I'VE SEEN ONE
FOR FIFTY YEARS

In the summer of 1970, I decided to take some time off. I was hoping to just relax, go to the beach, and hopefully catch a few ball games. I intended to live the laid-back lifestyle of a single, young American man.

But then my cousin Dick Harano asked a favor. He was going to drive across the country with his mother and my mother to visit his sister, Mary Lain, who was living in San Jose, California. I immediately knew where he was going with his request, so I let him grovel a little. He was so desperate for additional adult company that he offered to do all the driving and pay for the gas and meals, as long as I simply went with them. Being the kind soul that I am, I relented, and in August, we packed up his car on our trip. Little did I know that this journey would lead to the second most important decision in my life, the first being the acceptance of Jesus into my life. Another "God thing" occurred. Our drive across the country to the San Francisco Bay Area was uneventful, and I even relented and shared some of the driving. When we arrived, I met with Dick's sister, Mary Lain, while he took our mothers somewhere. With her was another woman, one of her co-workers from their company, Fairchild Electronics, located in Mountain View.

I fell in love at first sight. Her name was Kyung Ja Woo, but everyone just called her "KJ". She was an immigrant from Korea, sponsored by an American family of an executive of the plant. She was petite with a beautiful smile. That evening, our entire family was going sightseeing in San Francisco, and Mary had invited KJ to join us. Mary had intended to introduce KJ to her brother, but I met her first. Let's just say I was locked in on my mission.

I cannot remember much about the city of San Francisco, but my first evening with KJ was unforgettable. That evening we went to the usual sites, Fisherman's Wharf and some of the evening clubs. By the end of the evening, KJ and I were holding hands. Although my mother walked with us all night, I got KJ's phone number by the time we said goodbye. We remained in the San Jose area for seven days, and I took KJ out five of the nights there.

On our first date, I knew I had to show my best, and I borrowed Dick's car to take KJ out. When I pulled up to the sponsor family's house in an upscale suburb of Saratoga, my nervousness intensified. When I saw a grand fountain outside their house, I knew I was in trouble, coming from a lower middle-class family in the inner city. When I rang the front door, Mr. John Thomas, KJ's sponsor, opened the door. I was exactly on time, but he said KJ was not ready yet, so he invited me in. Mr. Thomas had two teenage daughters, Debbie and Devon, who had given KJ advice on dating in America. Namely, the girls recommended that she shouldn't be on time, and instead, make the guy wait. Since I wanted to get on the good side of the family patriarch, I initiated a conversation and asked where he was from. As it turned out, he was originally from Evanston, Illinois, a suburb of Chicago, which opened the floodgates to a discussion of the Bears, Blackhawks, and Bulls. In the meantime, KJ came out, but I didn't want to offend Mr. Thomas by cutting the conversation short. An hour and a half later, KJ and I finally left. The rest of the week was like a dream for me. Her radiant smile became my sunshine. She was very kind in our conversations. She exuded a special aura of beauty, both outer and inner. When she held my hand, I felt ten feet tall. After such a miraculous week, I knew I had to go all out trying to win her over.

After we returned to Chicago, I was determined to keep our connection thriving. I mailed her letters, made expensive long distance phone calls, and even sent her some special homemade 8-track sound tapes with a collection of romantic songs, such as "Misty", by Johnny Mathis. I cannot remember what we talked about on our long phone calls, but I loved just to hear her voice. I put in a lot of effort to get her attention, and had to do a lot of convincing. I invited her to Chicago to meet my parents during the Christmas holidays. By April, I had proposed to her, although it had to be over the phone.

Our wedding photo. Chicago, IL. 1971

In June, we were married. We had a special reception in Chicago, and then a beautiful honeymoon in Hawaii. God gave me the perfect mate that complimented my own strengths and weaknesses. I have never been a salesman, but I tell people that somehow convincing KJ to marry me was the best sales job I ever did.

When I met her, I was 26 years old, unemployed, and living with my parents in the Windy City. For her to accept my proposal and leave a rich family, a job at Silicon Valley, in sunny California? Furthermore, our different ancestries should have even been a dealbreaker, because I was of Japanese origin and she of Korean. Due to the animosity between the Japanese and Korean people, the odds of a Japanese man marrying a Korean woman, on a scale of zero to one hundred, was a negative fifty. My luck could only have been a "God Thing", and I have praised God every day since the day she became Mrs. Kyung Ja Mukoyama. Asking her to marry me was the second-most important decision in my life.

The first seven months of our marriage, we lived with my parents in their home. We all learned how to live in harmony together in the same house, and KJ and my mother get all the credit for that. My mother was just happy to see me get married, and KJ and my mother's bond turned out to be a happy blessing as well. Korean culture is quite similar to Japanese, so KJ adhered to the common cultural cornerstones. She honored my parents as respectful elders, and she cooked for all of us. Since KJ did not grow up cooking or eating American food, it was a time of learning how to prepare different dishes, especially some of my favorites. My Dad and I, typical guys, were clueless. Meanwhile, I had gotten a job with a very prestigious Japanese import/export firm, Mitsui & Co. I also joined the Army Reserves in the 85th Training Division, an Infantry training division.

Eventually, we bought a home in Skokie, a close Chicago suburb, and KJ started working at a photo processing plant which was owned by a Japanese American family and had many Asian employees. The owners were members of our church, and I had taught their daughter in Sunday school. The job was intended to be only temporary, until we had children, but turned out to be an eight-year employment. We tried to have children, and had a lot of fun trying, but with no success. But we knew that God wanted us to have a family. We contacted a number of different agencies for adoption, and when we found out that there were many orphaned children in Korea, we instantly knew what we had to do. Our adoption journey was twice as long as the natural nine months of pregnancy, but the wait was well worth it. It took us over 18 months, from the initial application to the joyous day our first child physically joined our family. The process was purposefully detailed and arduous for the protection of the child and to provide, as much as possible, a loving home, capable parents, and a safe, enriching environment. My wife and I were separately and jointly interviewed by social workers from the adoption agency. We had to undergo physical examinations and provide financial statements, including our tax returns for five years. We were required to submit essays about why we wanted to be parents. We were asked to provide letters from our employers, banks, and brokerage accounts. Our neighbors and church friends were interviewed about us as potential parents and the agency visited our home and went through the entire house. As grueling and lengthy the process was, it gave us confidence in the agency we were working with.

Our patience was eventually rewarded. We were so blessed to adopt two children who would become siblings: first a daughter, Sumi, and then a son, Jae, a year later. Since the children were coming from Korea, my wife's mother, Soo Soon Chun, who was living in Seoul, Korea, was able to visit the children at their foster home. KJ went to Korea to bring our daughter Sumi home, and they had a brief family reunion and celebration. Less than one year later, the agency contacted us with the wonderful news that they had another child for us, a boy. So, our family was complete with our son, Jae.

Adopting children was the happiest occasion in our married life, and not only for KJ and me — my parents also became proud grandparents. Our neighbors and church members shared in our joy, too, with baby showers and gifts. As we raised our children, we made sure that they learned they were adopted the moment we felt they could understand. A wonderful book that helped us accomplish this was *The Chosen Baby*, by Valentina Pavlova Wasson. Beautifully illustrated for children, the story describes a couple who prayed for a child, and God provided an adopted boy specially chosen for them. As a result, our daughter Sumi was so proud to tell anyone who would listen that she was specially chosen by God for our family. In fact, Lisa Rowe, her best friend from kindergarten through high school, came home one day and told her parents that she wanted to be adopted like Sumi, to be special like her.

Now came the responsibility of raising our children, which was guided by our family code. When our kids were kindergarten age, KJ and I decided that if anything happened to us, we wanted them to know our fundamental thoughts and values about life, family, and behavior. We therefore put together a short document entitled, "The Mukoyama Family Code." The title page had our Mukoyama Family Crest, and the title was hand-written both in English and in Japanese by my father. The first page listed "General Tenets," beginning with "Have faith in and be grateful to God." The second page contained "Daily Conduct," which was influenced by my Army daily training schedule, starting with "Make bed" and ending with "Go to sleep on time." On page three, "Prohibited Actions" were detailed, including behaviors such as "Do not brag about your accomplishments or possessions," and "Do not be discouraged by failure or disappointments — often such experiences in life give you lessons which will help you do better in the future." Although some psychologists will now advise never to make a list of "do nots," we were just doing the best we could as young parents without any Parenting 101 training or education. It always amazed me that the most important function in life of raising children was basically an OJT (on-the-job-training) experience. As The Mukoyama Family Code was a living document, the final page was entitled, "Addendum," where we would add on future guidance that we had omitted in the original pages.

Since our children could not read cursive nor long lists, we wrote each page in elementary print and colorful ink markers, then put each page in plastic sleeves in a three-ring binder. Our daughter's favorite color was purple, so the binder was a bright purple in color, and it became known as the "Purple Book."

When we introduced The Mukoyama Family Code, we made it a surprise occasion with a special dinner and ice cream dessert. We presented each child with their own copy. Over the years, there was an unexpected benefit. The kids occasionally failed to follow the code, but I didn't have to get into mad, scolding mode. Instead, we would instruct them, "Get out the Purple Book." We would then point out the value that was not honored, and they couldn't say, "You never told me." Yet in the Addendum, we added a disclaimer, "Always remember that Mama and Papa love you, even when we get mad when you break these rules." The Mukoyama Family Code became our structure for parenting, keeping both our children and us responsible.[3]

Over the years, my wife has demonstrated countless ways I am blessed to have her as a partner on my life journey. She has been an incredible wife, mother, and daughter-in-law. Three-generation homes are a prevalent tradition in Asian culture. The eldest son is supposed to be responsible for the parents in their old age, and my brother John had fulfilled that by buying a two-flat in Chicago and living there with our parents. But when their daughter, Lori, was born, he and his wife, Sharon, decided to move to the suburbs for the better education available. My parents were too old to get a mortgage on their own and didn't have the funds to do so.

Having been raised in an Asian culture, KJ immediately agreed to build an addition on our home in Skokie for my parents. Our three-bedroom ranch home soon doubled in size. We added a two-story addition, with a family room, fireplace, and two-car garage on the first floor, then three bedrooms and a bath on the second floor. My parents had a comfortable living space, one that was private but still closely connected.

A decade later, when our children were almost ready to start preschool, my wife and mother ganged up on me. "We need to move to a better school district for the sake of our children," the women argued. I might have countered with sending them to a private school, but I had little chance against that powerful team. Some may believe in the myth that meek Asian women walk three steps behind their husbands, but the Asian family is actually a matriarchy. Asian women understand that man is just a big blob of ego. So, they avoid embarrassing their husbands in public, and behind closed doors,

---

3 A full copy of the document is included as Appendix B.

the women are in control. Asian men can deny it, but it is what it is. I relinquished, and we soon moved to another home. So, for 22 years, my parents lived with us until they passed away. KJ was their caregiver until the end.

Reflecting on my wife's angelic qualities in this book would not be enough, and I try to cherish her each and every day. Around 2005, I wrote a short essay about KJ entitled, *What Is an Angel?* [4] At that point, we had been married for 35 years, and I witnessed her compassion and empathy every day. As I said in my essay, "I thank the Good Lord at the beginning of every day for honoring me with such a wonderful life partner. And it was clearly His hand that brought us together. How else can one explain two people, from two completely different cultures and continents and two different personalities meeting halfway across the continental United States?"[5] Our love story could only have been part of God's plan.

KJ's giving nature is one of her most honorable qualities. "Every day I marvel at her sensitivity to others and her attitude in virtually everything she does — from getting up at 5:30 AM every morning to prepare tea and a breakfast snack for me to planning every lunch and dinner meal so that I receive a nutritious and appetizing experience." But I am not nearly the only person KJ has served.

In addition to my bride's spiritual gift of serving others, KJ has a tenacity that belies her calm outer appearance. When she came to America, her goal was to get a college degree, an objective that was impossible if she stayed in Korea. Without reservation, KJ put that dream on hold when we started our family, but she never gave it up. By the time our children were in middle school, they no longer

---

4 A full copy of the document is included as Appendix C.

5 Quote from *What is An Angel*, 2005, in Appendix C.

needed her to be at home. She enrolled in a community college. By that time, my Army career had ended, so it was time for me to step up in our marriage. KJ had sacrificed her dreams for the sake of our family, so I understood supporting her would be the most important way to demonstrate my love.

I discovered that acts of service were KJ's love language, so I worked earnestly with her to help her achieve her academic goals. Since English was not KJ's primary language, reading took her three times longer than normal to comprehend her textbooks and assignments. It was often difficult for her to take notes. The solution was clear to me. She took a microcassette tape recorder to all her classes and recorded the lectures. Then, she would bring me the tapes, and I would sit at my computer with headphones and type her notes. I organized them and helped by adding highlights. I also typed all of her papers, following the APA style. I became her tutor and helped her prepare for tests. She did all the work, took the tests, and graduated Summa Cum Laude with a bachelor's in Psychology. She then got her master's degree in Gerontology, the study of aging. She was invited to make a poster presentation to the Gerontological Society of America National Convention in San Diego, and had a paper published on her master's thesis in the *Hallym International Journal of Aging* in 2005. I was incredibly proud of her, and our marriage was strengthened through our newfound scholastic partnership.

KJ then decided to enter the workforce again, but with the passion and purpose to serve others. As a wife, mother, and daughter-in-law, she had worked hard for decades, without a paycheck. Now she could put that education to good use and for the next 16 years, she worked at a rehab center with mostly senior residents. Her service qualities were clearly evident in her work, as she had arranged for special recognition for residents whose life stories would have gone unnoticed. Her patients always responded to her presence and touch with warm looks of affection, contentment, and appreciation. After retirement, she volunteered to serve a Japanese widow in her 90s. This widow of an Air Force Veteran has become part of our extended family and another manifestation of KJ's dedication and altruism.

I have not always been the best husband, as I've let anger get the best of me. Yet KJ has always endured through these hardships and responded with forgiveness. I have never employed any physical force, but as the Bible states, the tongue can be worse than a two-edged sword. With a lot of prayer and repentance and guidance from the Holy Spirit, I have improved. Do I still fall off the wagon occasionally? Regretfully, yes. Having said that, we recently celebrated our 51st wedding anniversary, so I guess KJ thinks I'm a keeper.

I could not have achieved the success I have reached in life without my wife's patience, grace, advice, and love. What is an Angel? You just met one.

CHAPTER 15
THE CITIZEN SOLDIER: UNAPPRECIATED
MASTER JUGGLERS

As with most active-duty soldiers and civilians, I had no clue about the Reserve components before personally joining. The Army has three major components: the Active component (AC), made up of personnel who are on duty 24/7; and the Army National Guard and the Reserves, referred to as the Reserve Components (RC). The RC are America's force of "part-time warriors." They are trained and ready to engage when called upon, but those serving in our Reserves rarely receive the recognition they deserve.

The Army Reserves (USAR) are a national asset, made up of part-time soldiers that are utilized and employed by the federal government. They are committed to a minimum of one weekend per month and a two-week annual training commitment normally at an active-duty post. As a part of the Department of the Army, they report to the Army chain-of-command and ultimately to the President of the United States. If the active Army Commander orders the Reserve unit commander in their area of command to jump, the response is "How high?" The Reservists are held to the same professional standards as the Active Component for promotions and assignments. The Army National Guard (ARNG) are also part-timers, but as part of a state organization, they report to the

Governor of their respective State. If the Army commander wants a National Guard unit in their area of operations to do something, the Guard commander can ignore such requests, because it is not an order. Their members are not held to the same standards as the AC, especially when it comes to promotion criteria. For example, an ARNG officer can be a Lieutenant Colonel one day and be appointed the State Adjutant General and become a Guard Major General. This would be impossible in the Army Reserves.

I need to emphasize two points here. First, this is not to cast any aspersions on the patriotism or dedication of the Army National Guard soldiers. Some of the most competent and professional soldiers I came across in my three-decade career were National Guard members. Second, for full disclosure, after five years as a Regular Army officer, I spent the next 27 years as an Army Reserve officer. I am admittedly biased. The military professional ethic requires soldiers to be non-political. Since the National Guard is a job program for the states, although the vast majority of their funding comes from the Federal Government, it is by nature political. That is not a criticism, just a fact of life. Although there are Army Reservists in every state and territory of our nation, they have no political clout, especially when compared to the Guard.

All Reserve component soldiers deserve the undying gratitude of our citizens. Yet, the roles of both the Reserve components, Reserves and National Guard, are both often unknown and underappreciated by the general public. That was certainly the case for me when I came off of active duty.

I joined the Reserves back in 1970. Back then, the Reserve components were rarely mobilized. But since then, the nature of the world and warfare has changed dramatically, with budget realities requiring a reduction in our Armed Forces personnel, resulting in a higher demand for the mobilization of RC service members.

To enlist in either of the National Guard or the Army Reserves is a selfless act of courage. Both the National Guard and Army Reserves are subject to being called to active duty to deploy to combat zones at a moment's notice. I fully recognize that they are receiving pay for their service, but they could be earning similar, if not higher, pay from part-time private industry jobs. The members of the Guard are additionally subject to be called up for state special emergencies unrelated to their military mission, and they always perform magnificently in those roles.

When National Guard units are mobilized or called to active duty, they are "federalized," which means they are officially under control of the Federal government and not the State. Fast forward 50 years, it is now common, not uncommon, for Reservists and Guard personnel to have been deployed overseas multiple times. The men and women in our Reserves are not only risking their lives when mobilized, but they are putting their civilian careers and relationships in jeopardy. They are working two simultaneous jobs because they have full-time jobs that are the main source of livelihood for their families.

Federal laws require that civilian employers keep jobs available for returning soldiers from active duty, but many deployed members return to lesser or different jobs, or their employers simply ignore the laws. Career advancement opportunities can also be lost. In addition, the family separation creates both disruptions and stress in romantic relationships, marriages, and bonds with their children. What causes these individuals to voluntarily remain dedicated to their respective branches of the Armed Forces? I believe that their dedication is based on their universal qualities of patriotism, selfless service, and shared values of the military culture. This doesn't occur by accident, but rather requires great leaders who exhibit **Care**, **Balance**, and **Example**. I have been blessed to have had such leadership and would like to introduce some special individuals to you. These four men mentored me at critical points in my Army career and taught me invaluable lessons. They were all highly successful in the civilian world yet dedicated numerous hours to serving our nation in uniform, first on active duty and then in the Army Reserves. The only aspect these mentors all have in common is that they personally educated, equipped, and enabled me to be the officer that I became.

First, Major General William P. Levine commanded the Army Reserves 85th Division (Training), and he demonstrated how a caring leader positively impacts his troops. He was a Jewish World War II Veteran, who participated in the Normandy landings and the liberation of the German death camps. He was the Commanding General when I joined the 85th Division, and I witnessed how he was loved by the soldiers because of his considerate leadership and caring nature. Because of the kindness he showed his troops, the support was reciprocated. General Levine emphasized the importance of our families, and when his first wife passed away during his command, the soldiers throughout the Division mourned for his loss with him. He was blessed to remarry later to a wonderful lady, Rhoda, who was his life partner until the end. The General took a liking to me as a junior officer and followed my career. The Levines were very gracious towards me and my family. In active duty, soldiers were focused on surviving every day on the frontlines, so they focused on the mission in front of them during deployment. But within the Reserves, I discovered the reality of the Reserves' sacrifice. Familial bonds were regularly put on pause for the sake of the job. Although the Reserves required parents to miss family dinners, weekend sports, and date nights, General Levine taught me that a good leader ensured that these sacrifices were recognized and rewarded.

The second notable mentor was another commander of the 85th, and later the first Army Reserve General to be appointed as a Deputy Commanding General for Reserve components at the Fourth United States Army headquartered at Fort Sheridan, Illinois. His name was Major General Wilber James Bunting. He truly saw my potential and helped me harness it. When I joined the 85th Division, General Bunting was assigned to the Division G-3 (Operations and Training) Section as the Division G-3 Officer. General Bunting and his wife, Carol, took us under their wings and invited us to their home. We got to know his family, including his mother. General Bunting had served with the 82nd Airborne on active duty and he gave me invaluable lessons regarding my military career. But equally important, he taught me the importance of maintaining a balance between my military and civilian careers.

Serving as both a great friend and example to look up to, Major General Richard E. Stearney was my Commanding General at the 86th Army Reserve Command (ARCOM). He gave me both invaluable companionship and career opportunities. My joining the 86th ARCOM was not only a blessing that resulted in a life-long friendship with General Stearney and his gracious wife, Jeanette, but rounded out the professional military qualifications that led to my final assignment. The 86th was one of the finest and most complex Army Reserve Commands in the nation, stretching virtually from coast to coast and involving nearly twenty major commands. I had the honor of being selected as Stearney's Deputy Commanding General, and was promoted to Brigadier, or one-star, General. The breadth of the command was extremely wide to include combat, combat service, and combat service support units. Its mission included a Special Forces Group, Artillery Group, Aviation Group, Engineer Command, Psychological Operations Command, Civil Affairs Command, Communications & Electronic Warfare Intelligence (CEWI) Battalion, and a General Hospital Command. Having been an Infantry and armor officer with mainly training experience, I had to be a quick learner, and Dick Stearney was just the tutor I needed. This assignment expanded my military knowledge and undoubtedly prepared me for all of my future experiences. The four years that I served as his Deputy was the impetus for the Army to consider me qualified for my next assignment as the Commanding General of the 70th Division (Training), headquartered in Livonia, Michigan.

My time with the 86th ARCOM resulted in numerous friendships, which have blossomed over the three decades since I left the ARCOM. I cultivated a very special relationship with our ARCOM Chaplain, Colonel Patrick Boyle, a Jesuit priest who taught ethics at the St. Mary's of the Lake Seminary, in Mundelein, Illinois. Father Patrick had been an Army Chaplain with the 1st Cavalry Division in Vietnam and was awarded two Silver Stars. Through our years working together, I never learned how he had earned them. But 50 years after his service in Vietnam, I attended the farewell retirement dinner for him, along with numerous members of his unit. These men were witnesses to his valor at a Fire Support Base that was under a tremendous siege with numerous casualties. Fr. Boyle could have evacuated prior to the siege but insisted that he stay with his sheep and provide comfort and support throughout the devastating ordeal. I had always respected him as a man of faith, but I was in awe of his heroic humility. Fr. Pat and I became close during my four years at the ARCOM because of my emphasis on faith to our soldiers. In fact, he often joked that I was more Catholic than some who attended his homilies.

My fourth and final example of leadership was 4-Star General Frederick M. Franks, Jr., who was the Commanding General of Training and Doctrine Command (TRADOC) at Fort Monroe, Virginia. I served in my final assignment as his Deputy Commanding General for Reserve Affairs. Although he was not a Reservist, General Franks commanded with a "one-Army" mindset, meaning he was always inclusive and mindful of his RC soldiers. General Franks was an Armor officer in Vietnam, where he was wounded and lost part of his foot. He was a West Point graduate and remained on active duty, attaining one of the highest command positions in the Army at TRADOC. General Franks understood the Reserve Components contributed tremendously to the success of our military as a whole. Personally, he was also one of my strongest supporters. When I had some run-ins with the Army brass at the Pentagon, General Franks' support was steadfast. On my final two annual Officer Efficiency Reports (OERs), General Franks maxed my scores and recommended that I be considered for Chief of Army Reserves. General Franks was a man of courage and integrity, and an active component guy who truly lived the one-Army concept.

I was blessed to have worked with experienced Reserve leaders, who not only demonstrated how to balance civilian, family, and Army responsibilities, but also played critical roles in the development of my Army career and leadership. As I hope you have seen, it takes a very special patriot to be a Reservist or National Guard member. I liken it to a juggler trying to balance three balls simultaneously — their personal/family relationships; civilian careers; and military duties. It isn't easy, but if you can do it, the rewards in personal satisfaction and sense of service are immeasurable. Please keep these brave members of our Reserve Components of all Armed Services in your prayers.

CHAPTER 16
CALLED TO THE COLORS AGAIN: BLAZER SHARP

My near quarter-century of service in the Army Reserves was marked with challenging, yet fulfilling, assignments. I commanded at every level: company, battalion, brigade, and division. In every instance, I had great Non-Commissioned Officers (NCOs) who luckily made me look good, and mentors who gave me challenging assignments to grow and didn't hand me my head on a platter when I screwed up. These individuals gave me the opportunity to reach my fullest potential.

My leadership philosophy also empowered me to rise through the ranks. By leading with care and balance, I set out to be an example for my troops. I prioritized visiting my troops and getting to know them personally. Even for the branches I never served in, I made an effort to understand the role they played. Time management is critical for all Reservists, but especially important for those of us in leadership positions. I was never consumed with rank or involved in Army politics; I just focused on being the best officer I could be, while still being an active father and family man. Although balancing my assignments, training, and personal life was a challenge, I understood that this juggle was necessary to fulfill my duty and set an example for my troops.

I attended numerous annual training assignments, where I served for two weeks on active duty throughout the nation, at such places as Fort Ord, California; Fort Bliss, Texas; Fort Riley, Kansas; Fort Sam Houston, Texas; Fort McCoy, Wisconsin, Fort Knox, Kentucky; Fort Benning, Georgia; and Fort Monroe, Virginia. Conspicuously absent from the list was the Pentagon. I avoided the Pentagon like the plague. I wanted to be with soldiers, rather than be a "desk jockey."

When I was promoted to Brigadier General (one-star), I was the youngest General in the Army at the time at the age of 42 years old. Subsequently, I became the youngest Major General (two-star) three years later, at 45 years old. When I took command of the 70th Division, I was the first Asian-American in the history of the U.S. Army to command a division.

The vast majority of my years in the Army Reserves was spent in individual training. On active duty, I had commanded an Advanced Individual Training (AIT) company training Infantry soldiers. I served 18 of my 25 Reserve years in Training Divisions, which are unique to the Army Reserve, as no other service had them. In times of mobilization, when we are called to active duty, the wartime mission was to proceed to an active Army installation responsible for training recruits in advanced branch schooling subjects. For example, we trained Field Artillery soldiers at Fort Bliss, Texas; Aviation soldiers at Fort Rucker, Alabama; Armor soldiers at Fort Knox, Kentucky, and Infantry soldiers at Fort Benning, Georgia. The Training Divisions were a cadre unit with subject matter instructors and drill instructors who instilled the fear of God into the recruits. The concept was that upon reporting for duty to the active-duty posts, the active-duty soldiers already there could be released to join warfighting units. The concept had never been tested — until Desert

Storm.

In 1990, due to my extensive experience with individual training both on active duty and in the Army Reserves, I was selected to command the 70th Division (Training), whose mobilization station was Fort Benning, Georgia. The nickname of the 70th Division was the "Trailblazers," which referred to their original unit members recruited from the Northwest Territory. In true Hackworth fashion, I created a unit greeting which was "Train Tough" with the response, "Blazer Sharp." No other training division had a special greeting and response. This unique maxim gave us a clear focus: train as hard as we could and stay sharp. Like Hackworth's nicknames did for our troops in Vietnam, this salute and motto boosted morale by giving the soldiers a special source of pride and purpose.

One of the first things I did at the 70th Division was visit and introduce myself to Major General Mike Spiegelmire, the Commanding General at Fort Benning. Since I had done my basic officer and Airborne training at Fort Benning, my return was like homecoming for me. When I walked into his office with an Infantry Division Combat Patch on my right shoulder and a Combat Infantryman's Badge, my credentials were established.

"I had the best training division in the Army, but I don't expect you to take my word for it. Observe my units for the next six months, and if they perform to or higher than expected standards, I request that my soldiers and I are given more missions at Fort Benning," I immediately informed Mike.

The Actives were reluctant to turn over the training to Reserves, which I understood, having been on their side of the fence in my career. But it would be a win-win for both of us. His soldiers would get well-deserved time off to be with their families. Although basic training is considered stateside duty, the Drill Instructors hardly ever see their families because they are with recruits 24/7. Additionally, my soldiers would benefit from actually performing their mobilization mission with recruits. Mike agreed, and he watched. My soldiers performed well. We had proven ourselves, and we were granted more missions.

MG Mukoyama visiting troops in the field. Livonia, MI. 1990

I also began writing regular letters to my soldiers, an initiative called *Mukoyama Missives.*[6] As I wrote in one of these letters, "The Purpose of these periodic letters is to convey information of critical interest to the men and women of the total 70th Division team — reservists, civilians, AGR's and AC members... It has always been my policy as a commander to maximize communications up and down the chain-of-command. It is virtually impossible for me to lead effectively without feedback from those who must live with and carry

---

6   *Mukoyama Missives* can be found in Appendix D

out my directives." I felt it was critical that my soldiers heard directly from me about my expectations of them, my evaluation of their performance, and the current state of affairs regarding our division mission. Whenever I visited a unit, I would make it a point to ask what the last Missive talked about. Reserve component soldiers, who are simultaneously juggling three balls — family, job, and military obligations — deserve no surprises. I have attached some Missive examples in Appendix D.

Six months prior to the start of Desert Storm, I saw the writing on the wall and alerted my commanders and staff to prepare immediately for mobilization. After 28 years of Army experience in training recruits, I was prepared for the scenario. We reviewed all of the mobilization plans and updated all standard operating procedures (SOPs). But on the darker side, it also meant making sure all of our soldiers and their families had their personal legal documents in order, such as wills, powers of attorney, burial instructions, and pay allotments.

I then sent a mobilization plan to our higher headquarters, Training and Doctrine Command at Fort Monroe in Norfolk, Virginia. I had a detailed evaluation of each of my 16 battalions in Michigan and Indiana. I knew that it was unlikely that my entire division would be mobilized at the beginning of any war, so I sent them a prioritized list they should follow for any mobilization from 1 (the first) to 16 (the last). I warned them not to screw with the list. In fact, I was the only Commander who sent them such a list. Eventually, this list was deemed so useful that it became a requirement for the other divisions. Sure enough, only half of my division was called to the colors, but it was by far the greatest number of units mobilized compared to any other training division.

More evidence of the high esteem in which the 70th was

considered was a precedent-setting decision to assign two active-duty battalions to one of my Reserve brigades. That meant that the active-duty Lieutenant Colonel battalion commanders would have their efficiency reports written by my Reserve Colonel brigade commander. To place the future careers of active-duty officers in the hands of a reservist showed respect for our command and soldiers.

Reservists were being burdened with the anxiety of mobilization and their familial responsibilities, and as a leader, it was my responsibility to find a solution to relieve it. I ordered my staff to set up a toll-free 800-number family support hotline to be manned 24/7. Such lines were expensive to maintain, and I got some resistance, but I overruled all objections. I wanted our soldiers to be able to concentrate on their mission without having to worry about their families back home. We would take care of them. I also drafted brief letters on my 2-star Army stationery for each soldier's employer and significant other, thanking them for supporting their soldier. With 1,500 soldiers being mobilized, that meant there would be 3,000 notes to be signed. My staff said they could set this up with a computerized signature. But I rejected this. Since my name is James H. Mukoyama, Jr., it took time to sign each letter, but I owed this much to my soldiers and their families and employers. Months later, the husband of one of my cousins was making a business call on one of his customers in Indiana and saw one of my letters hanging proudly in the office of the president of the company.

Our soldiers performed magnificently. They reported for duty in less than 72 hours, as opposed to our initial mobilization plan that called for us to spend one week at our Reserve stations to prepare. The war ended so quickly that our division was only at Fort Benning for 60 days, but the Trailblazers had proved their mettle. Our performance apparently caught the attention of my superiors too. I was selected to be the Deputy Commanding General for the U.S. Army Reserves at TRADOC, the Army's Training & Doctrine Command at Fort Monroe, Virginia. It was a total surprise, as I still had two years remaining in my four-year tenure as the commander of the 70th, and I had not requested consideration for the position promotion when it became vacant. I spent the next four years at TRADOC and was responsible for all 12 of the Training Divisions.

My leadership experiences in the Army Reserves were extremely rewarding. It was an honor to be the youngest Brigadier and then Major General at the time, and even greater to be the first Asian American in the history of the U.S. Army to command a division. I utilized the lessons I had learned from every step of my education and career, from ROTC to Vietnam. I implemented strategies that I adopted from LTC Hackworth and put my leadership philosophy of **Care, Balance,** and **Example** to the test. But the greatest test of my morals and leadership was still yet to come.

CHAPTER 17
STANDING TALL: RETIRING UNDER A CLOUD

Throughout my life, I have held my leadership and morals to a high standard. I was blessed with a great family and great career, in the greatest country in the world, so every day, I tried to act in accordance with the virtues of my faith and country. But 25 years into my Army Career, my integrity was tested. My life's mission was at stake. One moral decision changed my entire life, but it also highlighted a cancer in my beloved Army.

After the Berlin Wall fell and the Soviet Union likewise fell apart, politicians in the early 90s were calling for a so-called "peace dividend"; that is, a reduction in the Defense Department to transfer funds to social services. With the "Evil Empire" disintegrating, some of that made sense, and the Army had a procedure to determine how to reduce its force structure called the Total Army Analysis (TAA). TAA follows a very logical, methodical process. First, you determine the threats worldwide that the United States is facing. Next, you determine an ideal force structure (number, types and location of units; number of tanks, artillery, fighting vehicles, helicopters and fixed wing aircraft) and the logistics required. You then determine the funding that you can reasonably anticipate receiving from Congress. Finally, you readjust your force structure to accommodate the budgetary and world-situation realities.

Two years earlier, I and a group of senior officers, NCOs, and senior civilian technicians who were Army Reservists had formed a nonprofit organization called The Army Reserve Association, or ARA. Our mission was to educate Congress and the public about the role of the Army Reserves vis-à-vis the State Army National Guards. I had been an Army Reservist for 25 years and observed how the Reserves constantly got the short end of the stick whenever reductions or asset redistribution were involved. The Army bragged about the "One Army Concept" which meant full integration of the actives, Reserves, and National Guard. But it was in fact a myth. The actives looked after themselves, the Reserves had no political clout, and the Guard had so much political clout via their Senators and Representatives in D.C. that even the Actives kowtowed to the wishes of the Guard. The National Guard lobby was so powerful their annual convention is held in Washington D.C. in October and in every Presidential Election year, the two candidates from the Democratic and Republican parties both attend. Yet, the Army Reserves were frequently left out of the equation in Washington.

Back to the force reduction situation — the Army had to reduce its forces. Instead of using the established TAA process, the top brass — Vice Chief of Staff of the Army, the Chief of Army Reserve, the Chief of the National Guard, and Presidents of the Reserve Officers Association (ROA) and the National Guard Association of the United States (NGAUS) — met in a hotel room away from the Pentagon and made a smoke-filled-room deal. In essence, it decimated Army Reserve Aviation by 80% and eliminated the two Reserve Special Forces Groups, and kept the two National Guard Groups. Neither decision was defendable in terms of military readiness, which in fact was significantly degraded. It was a purely political decision. A state governor justifiably needs emergency medical, communications and logistics assets for state emergencies, but what justification is there for Farsi-speaking demolition experts? The last time I checked there was not a military insurgency in the state of Illinois.

Within five minutes of the conclusion of the meeting, I knew exactly what happened. I had ARA members in the room. I discovered that they were going to have a press conference the next morning with all participants holding hands and singing kumbaya and touting that this was a monumental "One Army" agreement without dissent. The problem was, that was simply not true. Wearing my hat as President of the ARA, not as a Major General, I called the office of the Vice Chief of Staff and asked to speak to the Vice about delaying the proposed press conference. I was told I would get a call back. I never received it. It never happened. The next morning at 8:00 AM, the ARA issued a press release saying the so-called "Off-Site Agreement" did not have the support of the men and women of the Army Reserve Association due to the detrimental effect it would have on our nation's readiness.

This was not the first time that I had been brought to the attention of the brass in the Pentagon. A few years before, I had appeared in uniform as a panelist on the Oprah Winfrey Show on the subject of racism and Anti-Asian actions against Japanese Americans due to our poor economy versus the success of the Japanese. I had been invited by Oprah because I was the highest-ranking officer of Asian ancestry in our military at that time. The invitation was on very short notice, so I contacted my civilian employer's main office in New York and received approval for a personal day off. I then notified the office of the Chief of Army Reserves. Shortly thereafter, I received a call from the Chief of Public Affairs for the Army, a Brigadier (one-star) General. When he learned that the show was not a military subject, he said it would not be good for me to wear my uniform.

I take my responsibility as a role model for minorities very seriously, and I saw this as an opportunity to excel, not fail. So, I told him that I appreciated his call, but if this was his recommendation as a staff officer, unless there was an order from the Army Chief of Staff, I would wear my uniform. He said he would call me back. Later that afternoon, he called and said he could not tell me it was an order. Clearly, no one wanted to put their name on the blame line. I told him that I would wear my uniform. I later received a copy of a Memorandum for Record (MFR) for my personal records signed by the Chief of Public Affairs to the Army Chief of Staff accurately reflecting our conversation.

I went on the show the next day. At the beginning of the show Oprah introduced me as follows: "Today we have with us Army Major General James Mukoyama, the highest ranking Asian-American today in our Armed Forces." The camera then switched to me. Imagine how it would have looked if I was sitting there in a civilian suit! Instead, I was in my Army Dress Green uniform with my ribbons and Combat Infantryman and Parachutist badges.

The very first words out of my mouth were, "Oprah, today the opinions that I express are solely that of Jim Mukoyama, not that of the Department of the Army or the Department of Defense. Having said that, I am proud to be a soldier in the United States Army, and we have equal opportunity in this country."

Since I had fully prepared, my appearance was a resounding success. I even received a letter of commendation from Secretary of Defense Dick Cheney. That letter arrived in my personal file and, in essence, canceled the MFR from the Public Affairs Chief.

By the way, I learned about the incredible influence of Oprah a couple months later. In the years before the heavy inspection of airport security, I was checking my baggage at the airport for a return trip home from an inspection trip to a Pennsylvania Army Reserve unit. The unit had given me a souvenir, a miniature jug of moonshine with my name on it and the unit designation and crest on the jug. When the scanner saw it in my bag, she asked what it was and I told her. She said it was a prohibited item and called her supervisor. He immediately came and took one look at me and said, "Aren't you the General who was on Oprah?" I said yes and he said, "Ah, go ahead," and waved me through.

Returning to the issue of the Off-Site Agreement, the Washington Post ran a front-page article with several quotes from the ARA news release. Soon thereafter, Army Reservists and Army Reserve civilian employees from throughout the country pleaded with me to testify. Someone had to stand up.

I remembered when I was a young officer in Vietnam vowing that if I ever got to a position of authority to protect my soldiers and nation I would do so, unlike the Generals and Admirals who failed to do so in Vietnam because they allowed themselves to be handcuffed by the elites in Washington. I remembered how Colonel Hackworth stood up and voiced his apprehensions about the Vietnam War. Although his military career deteriorated after sharing his concerns, he demonstrated that his troops and their well-being was far more important than himself. By standing up for his values, Colonel Hackworth chose to set a strong example of courageous and caring leadership. I had to decide whether I would follow his lead.

My opportunity to choose soon came. As President of the Army Reserve Association, I was invited, not subpoenaed, to testify before a Congressional subcommittee about the Off-Site Agreement. I could have excused myself by saying I had a conflict, or just declined, but as a leader within the Army Reserves, it was my responsibility. I needed to honor the Reservists I was serving with and leading. I had to tell the truth, even if no one else would. I accepted.

Prior to testifying, we had a family kitchen meeting where I presented the reality of the situation. I informed my family that testifying would most likely end my Army career, and our lives would change. KJ, as always, totally supported me. Our teenage children couldn't believe, based on my stellar military career, that doing the right thing would have negative results. I couldn't either, but I understood the consequences of doing both the right and the wrong thing. Although I was jeopardizing my life-long Army career, the safety and security of the Reserves was far more important.

KJ and I with Congressman Montgomery (D-MO). Washington, DC. 1990

So, I appeared before Congress telling the truth. I asserted that the sweeping overhaul of the Army Reserves would endanger soldiers' lives, degrade our military readiness, and waste our taxpayers' money. This off-site decision was haphazard, and the Guard's lack of preparedness would damage the U.S. Military's tactical missions and the grand strategy. Standing in front of Congress, I maintained that "politics should not be allowed to override military considerations."

Colonel Hackworth had also stood his ground to defend what was right, so he knew what consequences were coming next for me. In his nationally syndicated column, he tried to cover my six and protect my back by writing a column entitled, "Moral Cowardice Endangers Soldiers," a copy of which is in Appendix E[7]. He detailed my testimony before Congress and the background of the so-called "Off-Site" agreement.

James Mukoyama Sr. & COL David Hackworth (USA-Retired). Glenview, IL. 1990

Based upon his own experience with the Guard and Reserves, he agreed with my assessment that this haphazard decision was endangering troops by disregarding their tactical competence.

---

7   A full copy of this article can be found in Appendix E

Hackworth stressed that, "Courage is as much a part of soldiering as gunpowder. But having guts isn't just about charging the enemy. It's also about standing tall against wrongdoing and fighting for what's right." I greatly appreciated his cover and consensus, and Hackworth's own example of courage was one of the reasons I spoke out and told the truth.

It was unfortunately to no avail. One year later, I was history, and my career came to a screeching halt. Suddenly, there were no assignments available for me. Because I was so young, I still had five more years remaining before my mandatory retirement date (MRD). I got the message, and I retired.

Years later in a college application essay written by our daughter Sumi, entitled "Understanding," Sumi realized the impact of that testimony on my life. She recalled that life-changing kitchen conversation we had, and although she admits she did not understand the impact at the time, my testimony and subsequent retirement left a strong impact on her moral compass and understanding of the world. She remarked, "The political issue — the Active Army's decision to cut back on the Reserves — though important nationally, is not nearly as crucial as the personal ethical issue — standing up for what's right… It was not until [my father's] retirement dinner that I appreciated his bravery, and that his testimony before Congress may have been his most courageous act as a military man. My father chose the path that he thought was right; even though this action cost him his military career."[8] That essay means more to me than any military award I have received in my life.

By the way, when a General Officer retires with a 30-plus years career of honorable service, at the official retirement ceremony in

---

8   *Understanding* by Sumi Mukoyama can be found in Appendix F

front of the soldiers, there is normally a final honor awarded to the retiree, like getting a gold watch at a company retirement dinner. It is normally the next higher award that the retiree had received in their career. Since I had received the Legion of Merit, the next highest medal was the Distinguished Service Medal (DSM), the highest peacetime Army award and one that was not uncommon for General officers to receive. When I retired, I received a handshake. Remembering the Japanese phrase "shi gata ga nai," I understood that there was nothing I could be do about it. I just shrugged it off.

A few years later, the Chief of Army Reserves (CAR), Major General Max Baratz, was retiring, and he told me he had recommended that I receive the DSM upon my retirement, but he had never gotten an answer back. Normally, an award recommendation is either approved or disapproved. In this case, they simply buried it without action. There is actually a committee in the Pentagon that reviews "lost" award recommendations, so the CAR had his staff resubmit it. The Army leadership who had black-balled me had moved on, so when the committee reviewed the recommendation and my record, it was a no-brainer, and the award was approved. I was awarded the Distinguished Service Medal by General Jack Keane, the Vice Chief of Staff of the Army, in Washington, D.C. months later. This award was a nice token of solace, but as an American patriot and veteran, it can never fill the hole the Army left in my life.

Unbeknownst to me then, my career ending was actually a "God thing," but as often happens, it sure didn't feel like it at the time it was happening. As my daughter Sumi observed, "Even though my dad probably understood quite well what his testimony would do, I believe that deep down inside he was hoping for a storybook ending. Hidden within his heart, he wanted the Active Army to right its wrong and to continue his military career as a Commander at a new station. Underneath his realistic notions, his idealistic beliefs hid with quiet expectations." My daughter was quite an intuitive young woman, because she was right. At the time, I had devoted my life to the Army and was dedicated to serving, but I had to live up to my moral courage and compass. I had hoped that the Army would understand the crossroads this dilemma found me at, but I was disappointed. Nevertheless, I am now proud that I set an example of integrity for my daughter and others to follow.

CHAPTER 18
THE SECOND HALF: NEW DOORS OPENED

Bob Buford, a man of faith, wrote a book entitled "Halftime," aimed at Christian men, describing life in terms of a football game. Halftime is often a time of reflection on the past and future, emphasizing the latter, and the eternal question of life's purpose.

At 50 years old, I was throwing myself a personal pity party after seeing my 32-year Army career go down the drain. KJ always says she isn't as much a Christian as I am, but her way of life is often more aligned with faith than my actions. She had a woman-to-man meeting with me (remember that Asian woman thing?), and brought me back to earth.

"What do you have to be down about? You survived Vietnam. You have had success in business and the military. You have your health. You have a supportive family and, most importantly, married me!" KJ remarked. I couldn't argue with that wisdom. It was time to move on.

God began repositioning my life in preparation for my ultimate life mission. He addressed two major areas — spiritual and financial.

When we initially adopted our children, we joined Hillside Free Methodist Church in Evanston, Illinois. Originally, it was near our home, but the neighborhood around our Chicago church had deteriorated. Our car was even broken into during a service only half a block from the church. The congregation eventually moved to a safer neighborhood. Hillside was a very small church with maybe 40 attendees at our services, and the Sunday school consisted of our children and the pastor's children when we joined. We found Hillside through the yellow pages (the phone books at the time before the internet). We called on a Sunday afternoon, and the phone rang at the parsonage. The minister Reverend David Cooper answered, and he and his wife, Pam, invited us to come and see the beautiful church for ourselves. Even though its congregation of over 1,000 had split due to internal conflicts, there was still optimism for Hillside's future. The building and parsonage were paid for, and the remaining members were rock-solid believers who walked the talk. The church was firmly committed to the teachings of the Bible, and the members were very welcoming. KJ and I decided that this would be a lesson to our children that quality was more important than quantity, and we remained there for over a decade, becoming deeply involved and enjoying it immensely. But then our former ministers, Dave and Pam Cooper, answered another calling to minister a church out-of-state.

We joined a new church, Willow Creek Community Church, which created new and exciting experiences and concepts of church. Our teenage daughter Sumi suggested that we visit Willow Creek after attending with her best friend, Lisa Rowe, and her family. When Sumi stayed at Lisa's on the weekend, she would attend church with the Rowes — Pat, Linda, Maggie, and Lisa. Willow Creek was one of the original so-called "mega-churches." It was founded by a young pastor and a group of young adults, and conducted services in the Willow Creek movie theater in Palatine, Illinois. By the time we arrived, the church had grown to a congregation of 20,000 plus and purchased a huge tract of land in a northwestern suburb of

Chicago. It was a vibrant, evangelical non-denominational church with a mission to reach the unchurched and to transform them into fully committed followers of Christ. To me, Willow Creek's success resulted from four strong emphases. First, they were committed to the study and emphasis of the Holy Scriptures. Second, the strong spirit of volunteerism shined through, both in the generosity of giving and in the conduct of the services, maintenance of the buildings and grounds, the traffic guides in the parking lots, and the Sunday school programs. Third, the Youth programs were vibrant and continual, from infancy through high school. Willow became renowned for their program for infants through fifth grade, called Promiseland. Fourth, their emphasis on small groups fostered community and connection within the large church. It is impossible to know everyone in a multi-thousand-member congregation. But Willow addressed this and created connections by prioritizing small groups that met separately. I joined a men's small group and eventually became leader of a small group, which I still lead to this day. I was also co-director of a monthly Willow Men's Breakfast for over 15 years. Many different types of small groups were available: men only, women only, couples, and church ministry serving groups, to name just a few. We were hooked, and remained members for over 25 years.

My interest in men's ministry derived from another "God thing." Around the time we started to attend Willow, my brother John invited me and my son to attend a two-day outdoor event at Soldier Field in Chicago of a Christian group called Promise Keepers, or PK. It was the summer of 1996, the temperature was in the 90s, and 62,000 men filled the stadium. The purpose of PK was to get men together for worship, fellowship, and prayer. Guys could be guys. We wore T-shirts, jeans, shorts, caps, whatever. We batted beach balls in the stands around the stadium and issued chants and challenges to other sections. We'd yell, "We love Jesus, yes we do. We love Jesus, how about you?" then point to the opposite side to return the chant in a higher volume and point back. The day involved singing familiar hymns in lower musical octaves that guys can sing; speakers; and prayers for ourselves, our families, and our nation. The Promise Keepers intended to create unity and break down the walls of racial and social division between Christian denominations. The speakers addressed how men need to take responsibility to be leaders in their families. Leadership would empower them to be better husbands, fathers, and church members.

The Promise Keepers event in Chicago was an amalgamation of men of all races and ethnicities and all denominations, as well as many invited guests seeking fellowship and purpose. We had a boxed lunch both days, and I was impressed to see how organized and attentively attendees followed instructions to exit, pick up our lunches, eat outside on the grass, pick up their trash, and return on time for the afternoon session After the weekend, I asked my son Jae how he enjoyed the two days. He noted that it was a great father and son bonding experience, and I was grateful to have those moments with him. But even more so, he was impressed that with 62,000 guys in the hot summer sun, he didn't see one fight.

PK reached its zenith of popularity in the late 90s, with dozens of these rallies in sports stadiums and a historical National Day of Prayer in Washington, D.C., in 1997, with one million men showing up, including yours truly. This Day of Prayer was called "Stand in the Gap," and we met from noon until sundown and prayed for ourselves, our families, and our nation. There were no politicians invited because the Promise Keepers' faith and unity were far more meaningful than political gains. I attended as many PK events as possible over the next two years, traveling to Wisconsin, Minnesota,

Iowa, Missouri, Indiana, and Ohio. Although it has reduced in size and scope, these PK events still exist. The Promise Keepers, my men's small group, and my men's monthly breakfast planted the seed of men's ministry in my heart.

On the financial side, I got a job with a Japanese Import/Export firm, Mitsui & Co., in Chicago and worked there for five years. In Japan, being hired by Mitsui was as good as it could get. It was a company with an international reputation, and they had their pick of graduates from the best Japanese universities. I had hoped to promote good relations between the United States and Japan through business. I gained invaluable business experience by dealing in moving products throughout the world, international currencies, raw materials, manufacturing, and sales. But I had no future there. Since I was only of Japanese Ancestry, I experienced a form of reverse discrimination. I had my master's degree and more life and leadership experience through combat. I spoke a little Japanese and even had a Japanese work ethic. I arrived at the office at 7:00 AM and stayed until 6:00 PM, unlike my other non-Asian American co-workers who worked 9 to 5. But the Japanese office leaders looked down on me because I couldn't speak or read Japanese fluently. I didn't expect guarantees, but I expected equal opportunity to compete. I could have been there for 20 years, and the highest I might have become was a section manager. So, I offered my resignation and thanked them for the opportunity and experience.

When I was at Mitsui, I was reminded that my lifelong connection to the military did not end when I left active duty. While on the job at a headquarters in Chicago, I was sitting in a waiting room and picked up a *U.S. News & World Report* magazine on the coffee table. It had an article about Vietnam, so I picked it up and started reading it. A small photo showed a soldier being carried on a stretcher. Suddenly, I realized that the soldiers in the image were my men! I then saw that I was even standing in the background!

9th Infantry Division, Mekong Delta, Vietnam. 1969 - UPI Photo (Shunsuke Akatsuka)

I contacted the magazine and was given the information on how to get a copy from United Press International (UPI). One of their photographers was on the medical evacuation (medevac) helicopter I had called to pick up my wounded soldier. The soldier was Specialist Five (SP5) Barry Rabinowitz, and he was my radio-telephone operator (RTO). Among the highest casualty rates in the Infantry units were the RTOs and the junior officers. The enemy knew that if they took out our communications and leadership, their chances of victory immensely increased.

The photo was just a glimpse into our service and Barry's sacrifice, but it carried an even stronger meaning to me. I later learned that the UPI photo was selected to go around the world that day and appeared in all major newspapers, including the *Stars & Stripes*. This newspaper circulated to our service members serving at home and overseas. The word came back to the Battalion that Barry had been evacuated back to the States, which was usually a good sign of survival. Unfortunately, we later learned that Barry did not make it, and his name is on the Vietnam Wall Memorial in Washington, D.C. Exactly 50 years later, my wife and I were attending a reunion of the Hardcore Battalion in Tampa, Florida. At the general meeting, Infantryman Mike Art raised his hand and said he had something for "General Mook." He had saved the original *Stars & Stripes* newspaper with the photo prominently displayed for half a century, and he generously gave it to me. That is the depth of love and selflessness that bonds Veterans of all ranks, services, and eras together.

One of my platoon leaders, Carl Ohlson, took these efforts one step further. Carl is an art instructor, who made a beautiful painting rendition of that photo. Since I was the company commander, and in the picture, he offered me first dibs. Needless to say, it is hanging in a place of honor in the Mukoyama household. By the way, Carl also appeared in the far corner background of the photo, but he modestly opted to remove himself.

Throughout the nearly five years I worked at Mitsui, the Good Lord brought another person of faith into our lives, who became a life-long friend and mentor. His name was Kei Harada, a Nisei, who had served in the military as an interpreter and had served in Japan during the occupation and Korean War. There he met his wife, Reiko, a Nisei who was born in Seattle when her father, a Japanese businessman, was stationed in the States. Kei actually interviewed me for the job and was instrumental in my hiring. When it became evident that my future was limited there, he agreed with my decision to leave when I was still young enough to do so. Because Kei could read and write Japanese fluently, he had done well at Mitsui, but even he had reached his highest level possible as the Assistant General Affairs Department Manager. He was the exception that proved the rule of limited advancement for non-Japanese nationals, even in the United States affiliate.

A former university roommate and fellow Pershing Rifleman knew I was frustrated at Mitsui and offered me a job to join his operation on the floor of the newly formed Chicago Board Options Exchange. I didn't have one formal day of college education in finance, but he knew my abilities. My 30-plus-year career in the financial services industry began. On May 1, 1975, the securities industry went to a fully negotiated rate basis. This became known as "May Day" in the securities industry. Prior to that, there was a standard commission table that all firms followed. Now firms could charge whatever they wanted. The Discount Securities industry was born, and my former roommate and I started our own firm. We were very successful and sold our company in the late 1980s to a New York Stock Exchange firm. Although I did not become independently wealthy, I was able to make a significant return on my investment so that I had a comfortable nest egg to invest. I still had to work, and now had enough experience to qualify for a management position in the industry, so I worked as the Chicago branch manager for the New York firm that bought us out. In fact, my staying was a requirement for the deal. Years later, my former partner convinced me to join him as the Executive Vice President and Chief Compliance Officer in another firm. I neither sought nor received equity but negotiated a substantial severance package.

God had me on his potter's wheel, but I was still in the formative stage.

WISDOM

CHAPTER 19
SERVING: SURPRISE OPPORTUNITIES

As my life transitioned away from serving in the military, this gap was replaced by a new form of service — within the community. Between the ties I had in local Chicago and Veteran associations, I quickly found a new purpose in supporting people in need of resources or care. Within the next decade, I became involved in five significant serving opportunities.

When KJ completed her master's degree in Gerontology, she sought experiences to utilize her newly learned skills. After she saw an article about a hospice organization asking for volunteers in a nearby suburb, she quickly joined their six-week training program. Being a protective husband, I didn't want her to go alone at night, so I volunteered to drive her there, wait, and then drive her home. But I got sucked into the training, so we both qualified as patient-serving volunteers.

A hospice team comprises a doctor, a nurse, a certified nurse assistant, a therapist, a chaplain, and a patient volunteer. Hospice aims to provide palliative care, making the final days of a person's life the most comfortable, high-quality time possible with sensitive, caring service. Accordingly, hospice organizations are among the most loving and comforting anywhere. The patient volunteer is trained to provide minimal medical procedures such as changing diapers, Foley catheter bags, and administering oral medicines. The primary function was to provide socialization for the patient and a break for the primary caregiver. The unit of care is not just for the patient, but for the caregiver and their family members too.

After completing our training, KJ and I knew we would be split up because there was always a high demand for patient volunteers for the teams. One night, we had dinner with a close friend whose husband had passed away from cancer. When we mentioned our involvement in hospice, she commented that the most effective time of support was when a hospice husband and wife visited together because the wife would be comforting our friend while the husband would be with her husband. We immediately shared her insight with our Rainbow Hospice Patient Volunteer Coordinator.

"Have you ever assigned a couple to a hospice team?" we asked the coordinator. She responded, "You two are the first couple who had ever gone through the patient volunteer training together, so let's try it!"

It turned out to be a big success, especially for us. Being a patient volunteer is very stressful. Intellectually, you must understand the contract that the person you serve is dying, and you need not get too close as friends. But we are all human. The patient is often alone; their spouses may have passed away, or their family lived in another state. You become their surrogate family and bond with them. Being a team, KJ and I could lean on each other. Together, we talked about the rewarding benefits of service, lamented the hardships, and remembered those who had passed. The average hospice patient volunteer lasts 18 months, but we were blessed to serve for six and a half years.

One exceptional patient we had was a Vietnam Veteran named

John, an enlisted man in the Army. Whenever I would take him in his wheelchair around the facility, John would tell everyone that he had a General pushing him around. At a very young age, he had immigrated from Greece, with only his widowed mother, after his father died fighting in the Greek Civil War. His mother spoke little English and worked as a seamstress with Hart, Schaffner & Marx in Chicago for decades. She never remarried and devoted her life to raising her son.

Likewise, John dedicated his life to caring for his mother and never married. After John returned from Vietnam, he went to work for the federal government, and had recently retired and planned to reward his mother with a trip around the world, but he contracted cancer. John was a true fighter. He professed that when he beat his illness, he would become a hospice patient volunteer just like us. KJ and I had so many hospice patients during our six-and-a-half years with Rainbow that we had a policy of not attending funerals. But John was the exception. Between his personal story and fellow Vietnam combat history, our attachment to John moved us to attend. At the funeral, we met his mother, who was incredibly kind and gracious toward us. All of his family and friends knew who we were and welcomed "the General." The experience came with intense emotions but immense rewards.

If you have a friend or loved one diagnosed with a terminal illness with less than six months to live, run — do not walk — to the nearest hospice organization. Medicare pays for the service, and the hospice organization should not charge you for any medical services, equipment, or medicine. There are even hospices with specific and special programs for Veterans, such as our organization — Rainbow Hospice and Palliative Care. I was honored to be involved in the early stages of the program at Rainbow.

Heiwa Terrace, a senior citizen Section Eight subsidized housing apartment complex in Chicago, presented a unique opportunity for me to get involved and serve. The facility was initially funded and built by the Japanese American Service Committee (JASC), a nonprofit organization providing social services for aging Issei in our community. As the Japanese American population aged and many moved back to the West Coast, other ethnicities moved in, including many Korean elderly. I was invited to join the Board of Directors by one of the original members of the JASC, on account of my experience in the Chicago community, Japanese population, and hospice. I served for seven years, the final three as President. It was an introduction to 501(c)(3) non-profits, which became essential for my later endeavors. I learned about the government tax record requirements; the fundraising aspects of nonprofits; and the duties, responsibilities, and recruitment of Board members and volunteers. I developed a deep understanding of the nonprofit world, which I applied to all my other service opportunities. The pottery was taking shape.

A few years later, I received a call from the Department of Veterans Affairs, informing me I was nominated for a position on a committee at the VA Central Office (VACO) in Washington, D.C. For my third serving opportunity, I was a part of the Advisory Committee for Minority Veterans (ACMV), whose mission was to ensure that minority veterans received equal treatment from the VA. When I inquired who nominated me, the response was the White House. So, I accepted, and I ended up serving on the ACMV for five years, the last two as the Chairman.

Like most Vietnam Veterans, I didn't have a clue about the VA when I returned from the war. Society as a whole treated the Vietnam Veterans so poorly that many assumed there was a lack of resources.

They were already troubled enough, so they had no interest in dealing with more bureaucracy problems in the VA. The scuttlebutt about the quality of the VA services was not good. So although I had been slightly wounded, these minor injuries healed, and I knew many others who needed medical assistance. I didn't want to pile on the system, so I didn't enroll for health benefits. But nearly three decades after the end of the Vietnam War, the VA had thankfully improved dramatically.

Dinner with Secretary of Defense William Cohen, Janet Langhart, Sen Inouye, and Shiro Shiraga. Washington, DC. 1991

The VA has three major components: the Veterans Health Administration (VHA) that runs the hospitals and clinics; the Veterans Benefits Administration (VBA), that processes all of the benefit claims; and the National Cemetery Administration (NCA) that administers all of the National Veterans Cemeteries throughout the nation (except for Arlington National Cemetery, which the Department of Defense manages). During my tenure at the VA, I visited dozens of hospitals, benefit offices, and cemeteries, as well as conducted open town hall meetings for Veterans and their families. Although there are always exceptions with the largest medical system in the world with over 250,000 employees, I discovered that the VA service was generally excellent. Minority Veterans were treated equally. But the major problem I found was a dismal failure that affected all Veterans: lousy outreach. This lack of outreach was unfortunately especially true in the minority communities. There were sparse announcements or ads in the minority language newspapers, television, or radio stations. Veterans should have been informed of the programs available, ones that would have dramatically improved their quality of life.

To James H. Mukoyama Jr.
With best wishes,
Chairman Mukoyama —
Thanks for your leadership
Eric K. Shinseki
Secretary of Veterans Affairs

Advisory Committee for Minority Veterans, Department of Veteran Affairs.
Washington, DC. 2009

Every year, we submitted a report to the Secretary of the VA with specific recommendations. We received thanks and promises of action but did not see any significant progress. Year after year, we heard the same excuses and never saw any solutions. I, along with my Vice Chairman, a retired three-star Navy Admiral, and another member, a retired Army Colonel, submitted our resignations. The Secretary pleaded with me not to resign and promised things would change. I decided to see, and, sure enough, the threat made a difference. One of our major recommendations was approved and implemented, dramatically improving access to medical treatment for Alaskan Veterans not near VA facilities. This effort was majorly thanks to Committee member Nelson Angapak, a native Alaskan fellow Vietnam Veteran, who articulated the need and recommended a solution in great detail. Having accomplished the change, I withdrew my resignation and finished my term. However, I felt my efforts could be more valuable elsewhere, so at the end of my term I left my post at the VA.

The fourth serving opportunity consumed seven years of my life but was a true labor of love. The Nisei World War II Veterans are my heroes. If it weren't for them, Asian American service members — myself included — would not have had the opportunity to succeed. They paved a path for us by sacrificing their blood and demonstrating exceptional valor on the battlefield. I was invited by a Nisei Veteran member of our Chicago American Legion post to join the National Japanese American Memorial Foundation Board of Directors. Its mission was to build a memorial in Washington, D.C. to honor the patriotism of all Japanese Americans during World War II, despite the abrogation of their civil rights that put 120,000 persons of Japanese Ancestry into concentration camps. Their heroic service and character deserved to be preserved in our nation's Capital.

As we organized efforts to honor the Niseis' service, I received more education in nonprofit organizations — but at a national level. The Directors were all very successful in business, education, and government. Most had been heads of their departments or corporations. I didn't have a lot of financial connections for fundraising purposes, but I did have considerable influence among the Nisei Veterans simply by being a General and Infantryman in combat. So, I leveraged these connections to meet my obligations and help us accomplish the mission. For three years, I traveled nationwide on my own dime, giving speeches to Nisei Veteran organizations and Japanese American Citizens League (JACL) meetings. When I appeared at these events and in numerous newspaper articles, I simultaneously promoted the value of the Memorial and pitched for financial support.

National Japanese American Memorial Foundation Board. Washington, DC. 1999

While working on the Memorial, I had the opportunity to contribute to different projects within this effort. I helped to edit and publish a book detailing the Memorial's journey from securing the Congressional Law authorizing the Memorial to its dedication in 2000, titled *Patriotism, Perseverance, and Posterity*. I volunteered to chair one significant committee.

Our committee had to identify and list the names of the soldiers who served and died in the segregated World War II Japanese

174

American Army units. This procedure was extremely painstaking because the Army did not maintain records in those days according to race. Also, records were not computerized, so the initial data had to be gathered the old-fashioned way, by researching physical archives and documents. We collected information from the national archives in D.C., and most helpful were records and books from the Nisei Veterans organizations, especially in Hawaii and the West Coast. Unfortunately, there were inaccuracies in all of the lists from published books and the Veterans organizations, such as the same person being listed twice with different spellings, or reports of some being killed in action (KIA) but later discovered to be released prisoners of war (POWs). Hence, we had to do tedious work to ensure all names were correct. We finally devised a list of over 800 verified names titled *Died In Service.* We posted this list online and in Japanese American newspapers throughout the country for six months, requesting any corrections.

I was adamant that all names must meet the minimum criteria to be listed. Being carved in granite, we couldn't just erase errors. Twenty years after the dedication of the Memorial, there have been no reported inaccuracies. During the compiling of the list, there was one incident that arose. The units originally had Caucasian officers, many of whom died in combat. There was some objection to listing their names because they were not of Japanese ancestry. Being a combat Veteran, I immediately quashed such nonsense. This was a National *American* Memorial. To deny those soldiers this honor would have been reverse discrimination of the worst kind. I permitted no further discussion (it's sometimes helpful to be a General).

On the fundraising side, we had a deadline of seven years to raise enough finances to build the Memorial, or it would be canceled. Our goal was $9 million. The country was split into seven regions, and the Midwest was headed up by what was nicknamed "The Chicago Trio." It included Shiro Shiraga, a businessman; Mas Funai, an attorney; and yours truly. All of us lived in the Chicago area. Shiro and Mas had Japanese American connections and did the heavy lifting. I assisted with the Veterans organizations and presented the speaking engagements. As a Foundation, we exceeded our goal within the deadline with over $13 million collected, with the Midwest region collecting nearly $2.8 million. The Memorial was dedicated in November 2000.

Groundbreaking National Japanese American Memorial to Patriotism with General Shinseki. Washington, DC. 2001

My fifth significant serving opportunity was a collision of two of my greatest passions — spreading the word of God and supporting our brave Armed Forces. Retired Army Major General Joe Gray had served with me in the Army Reserves, and he was a very dear friend and a strong man of faith. We used to attend General Officer conferences and pray in our hotel rooms in the evening. After retiring from the Army, Joe became the Executive Director of the Military Ministry of CRU, formerly known as Campus Crusade for Christ. One day, Joe called me and asked if I knew anyone at the U.S. Naval Station Great Lakes. At that time, I had no contact with any of the leadership at Great Lakes. The CRU Military Ministry had a program to augment the command religious programs at the various services' basic training centers. They had active programs with the Army, Marines, and Air Force but had failed to get their foot in the door with the Navy.

Joe asked me if I could help, and I agreed to try. The Navy is very keen on military protocol, so when I called the Admiral's office at Great Lakes and identified myself as Retired Army Major General Mukoyama, I was immediately connected. I told the Admiral, whom I outranked, that we had a volunteer group offering to assist and augment the chaplains at Great Lakes. The Admiral agreed to a meeting, and the rest is history. I have since been a volunteer instructor at Great Lakes at the Recruit Training Chapel for 20 years. I tell the recruits that my blood now runs part Navy Blue. It's not crucial to me whether the recruits remain in the Navy as a career. I want them to have the same peace of mind and joy that my faith gave me during my time in service and combat. This objective supports the mission of the Recruit Training Command (RTC) to produce the best sailors possible for our nation.

I would not have been able to serve in these opportunities without the strong support and encouragement of my loving wife KJ. She saw the value of service, and her sacrifices enabled me to devote myself to these five causes. In one of my too infrequent times of recognizing her significance in my life, I wrote her a Mother's Day letter in 2003. I did a reprise letter nearly two decades later.

Both are attached in Appendix G.[9] I thanked her, as she was "always supportive of all of [our family's] endeavors, especially me, as I pursued my career in service of our nation. Never once did she complain that her load was heavy."

My education, Army career, and religion were the building blocks of my life, and these service opportunities were constructed out of them. And through each of these five endeavors, I gained enlightening lessons through my hands-on work. All these experiences covered in this section of the book equipped me to proceed into the fourth quarter prepared to apply them to my particular areas of knowledge.

---

9 Mother's Day letters from 2003 and 2020 can be found in Appendix G

CHAPTER 20
GENERAL'S MANTRA: AN UNLIKELY ARROW

I am a very grateful man. I have seen the blessings and opportunities God has afforded to me. At every chance I get, I thank the Lord and the special people around me. But as I go through daily encounters and experiences, *my mantra* exemplifies my optimism and gratitude.

In one of these special opportunities, I was interviewed by Jocko Willink — a retired U.S. Navy Seal Officer, highly-decorated Combat Veteran, author of the #1 New York Times bestseller book *Extreme Ownership*, and producer of the *Jocko Podcast*. My son Jae, who is into martial arts and physical fitness, is an avid follower of Jocko. He heard Jocko mention his admiration for COL Hackworth's famous books *About Face, Hazardous Duty,* and *Steel My Soldiers' Hearts*, the latter being about our Hardcore Battalion in Vietnam. Jocko had never met Hack, and Jae contacted Jocko, mentioning that I had served in the Hardcore under Hack. Jocko immediately reached out to me. When I got a speaking engagement in Los Angeles at an Army Reserve Command, Jocko with his assistant Echo Charles came up to the hotel I was staying at and recorded a podcast (No. 124). The interview lasted over three hours, and we discussed Hack's last book and how I got to know him.

Jocko also made an effort to learn more about my own life.

During that podcast, I, of course, mentioned my saying of "Every day is a great day! I have my faith, my family, and live in the finest country in the world." I only discovered later that Jocko extracted the seven-minute discussion in that podcast about my saying and made it a separate brief podcast entitled *The General's Mantra.*

Appearing on the *Jocko Podcast.* San Diego, CA. 2018

I say this mantra every day, every chance I get. The most common question we all get asked is, "How are you?" And my response is always the same. I repeat my mantra and mean every single word. I say it when I meet people, every time checking out at a store, talking to a bank representative, or a customer support system representative for online business.

Sometimes I get pushback about the United States being the "finest country in the world." I often hear, "How can you say that with all of the division in our country?"

"I've been around the block a few times," I always respond. As a kid, my chances of becoming a General in the United States Army were slim and next to none. But now I have seen an African American elected President of the United States and re-elected. I have seen African American Secretaries of State, both male and female, and an Asian American Chief of Staff of the Army. I have witnessed legislative improvements that have effectively ended real-estate redlining, poll taxes, segregation in the military, schools, and public areas, and bans on interracial marriages. What does that tell you? In my 79 years of life, I have witnessed tremendous improvements in our society for minorities. Do we still have room for improvement? Absolutely, but the United States is one of the largest multicultural nations in the world, and we are consistently pushing expectations to improve conditions.

In June 2020, I authored and published an article in a local newspaper entitled *Systemic Racism is Not Alive in America*. As the son and husband of non-white immigrants, I have seen that much of America's appeal and merit lies in its promise of opportunity. I decided to voice my experiences to contrast with the narrative media and educational institutions promoted. My Army Career proved to me that our society functions with equal opportunity. Promotion was based on ability, work ethic, and results, and we all received equal pay for equal work. The color of your skin and gender was irrelevant — we were all Army green.

As my article posits, "I have experienced prejudice in my life, but throughout, I received strength from my trust in God. I have also met and experienced people in my life who provided help, hope, and encouragement. The vast majority were not people of color but white. These experiences are, in fact, the systemic nature of our American society."

The article was published through the efforts of a fellow Vietnam Veterans of America Chapter 311 member, Greg Padovani. Greg is the Chairman of the Veterans Memorial Committee of Arlington Heights, Illinois, which is one of the most patriotic communities in our country. The Arlington Heights parades on Memorial

Day and the Fourth of July are legendary in size and community participation. Greg is dedicated to spreading the principles of Americanism throughout the Chicagoland area and has one of the largest rolodexes (contact lists, for younger readers) of Veterans and community organizations that I have ever seen. Equally supportive of our Veterans and First Responders is the Mayor of Arlington Heights, Tom Hayes, who is an Army Veteran himself and a West Point graduate.

I faced some harsh reverberations for voicing my truth. As a result of that article, I was "disinvited" to speak at a Fortune 500 Veteran Employee Resource Group because I was considered "too controversial." I will let you be the judge of the full article, which you can find in Appendix H[10]. As you know from elsewhere in this book, it is not the first time I have stood up, and Lord willing, it will not be the last. Through the wisdom of my age and faith, I am not afraid to share the truth. My absence of fear is a benefit of being an elder warrior and a person of faith.

Regarding faith, we are among the most tolerant nations in the world regarding freedom of religion. I take strong issue with those who try to restrict our religious freedoms. No phrase in the Constitution mentions the separation of church and state. That phrase emanated from a letter from President Jefferson in 1802 to the Danbury Baptist Association that referenced the Establishment and Free Exercise clauses of the First Amendment to reaffirm the right to free religious practices for American citizens. It has been misused for the agenda of those wanting to remove God from our lives. Our nation's founders clearly thought otherwise in our founding documents, when they stated "… fundamental rights are endowed

---

10 *Systemic Racism is Not Alive in America* can be found in Appendix H

on every human being by his or her Creator." Our Constitution refers to freedom "of" religion, not freedom "from" religion.

When I assist at the Recruit Training Center (RTC) Chapel at the Great Lakes Naval Training Center, followers of any faith have the opportunity to pray in their religious traditions. Areas are set aside for meditation and communing with whatever supreme being one selects. And no one is required to attend chapel. This is true freedom of individual rights.

Speaking of the Chapel, one of the strongest examples of faith in my life was my dear late friend, former Air Force Chaplain Dr. Jerry Hardwick. After the Navy approved the Military Ministry of CRU to support the command religious program, Jerry was appointed to implement the program and recruit volunteer instructors. He was the obvious selection because he started the highly successful CRU Military Ministry program at the Air Force Basic Military Training at Lackland Air Force Base in San Antonio, Texas. For over half a year, Jerry commuted between Chicago and his home in San Antonio to work with the area's RTC chaplains, churches, and divinity schools. When he was ready to unveil his *New Centurions* program that later became *Sailors for Christ*, he sent invitations to over two hundred Campus Crusade for Christ program participants to a special presentation at the Trinity International University near Great Lakes. I received an invitation because of my earlier involvement and had no intention of getting further involved, but as a courtesy, I decided to attend. A large conference room was set up with refreshments. The base chaplains attended. A total of two people, including me, showed up. After Jerry's presentation, he took full responsibility for the poor turnout and said that God wanted this program to succeed, so he would go back to the drawing board. When I saw his faith and dedication, I told him, "I'm in." That was in 2002. Two decades later, the program is alive and well, and thousands of United States Navy Sailors have benefitted. And I have been privileged to continue as a volunteer instructor since the beginning.

But Jerry's story goes beyond his lifelong service to the Lord. In early 2017 he received a prognosis of 6-12 months to live due to stage 4 cancer. Having ministered to hundreds of individuals with cancer over the years, he and his wife of four decades, Magdalena (Maggie), decided not to undergo traditional cancer medical treatments but to submit themselves to "Holy Spirit Radiation Treatments." Not only did they not reduce their national and international preaching engagements, but they also doubled down on Jerry's efforts to include conferences and churches throughout the United States, Central America, Spain, and Israel. His impactful sermons often went 1 to 2 hours. Jerry was also fluent in Spanish, as he had met his bride in Spain when he was in the Air Force. He also founded the New Centurions Evangelistic Association with outreach to military chaplaincies in Latin America.

Early on, Jerry decided to share his journey by writing a weekly email blog entitled *One Foot on a Banana Peel*. In late 2017, he and his wife visited the Great Lakes Naval Training Center, where numerous individuals came out to share their testimonies of how Maggie and Jerry had touched their lives through their ministry. KJ and I hosted them at our home, and I presented Jerry with a three-ring binder that cataloged every Banana Peel Blog written. This gesture was a token of my gratitude for the love, good work, and inspiration he modeled for me. I have included one of his essays in Appendix I[11] of this book. He continued posting his blog for nearly

---

11 *One Foot on A Banana Peel: The Power of Giving Thanks* — Appendix I.

19 months. The Good Lord called Jerry home on January 22, 2019. I have not met a stronger man of faith in my life, and I was honored to be his and Maggie's brother in Christ.

Once Jerry established the Sailors for Christ program at the RTC, the program was then organized and led by John McIntosh and his wife, Karen. They have nurtured the program well beyond the initial concept. Their beautiful home was within a stone's throw from the base, so they welcomed active-duty Navy personnel and their families stationed at Great Lakes and hosted evening and weekend activities. Over the past two decades, Karen has earned the affectionate nickname of "Mama Mac." She is the ultimate hostess, and John isn't so bad himself. They provide home-cooked meals, and their finished basement provides video games, music, and snacks. It is a home away from home and a great fellowship experience. As a result of their efforts, thousands of recruits and hundreds of active-duty personnel have either come to faith or renewed their spiritual commitments.

I have been blessed to be a Christian virtually since birth. With my baptism, my parents committed themselves to raising me according to His Word. But I do not force my beliefs on anyone. It is always the individual's personal decision. I strive every day to live my life according to God's commandments and be a living reflection of the love of our God.

I also am prepared to share my faith when the opportunity arises, when asked about my faith. People can be curious why I act the way I do or what my source of peace and joy is, and I always reflect on God's will and words.

Regarding my family, I am not only referring to my biological family but my dear friends and associates in life. I fully realize and appreciate that not everyone has loving parents and siblings. Having served in the military, I was exposed to various people who endured terrible life experiences. Almost nothing surprises me anymore. Upon hearing horror stories of abuse, tragedies, betrayals, and terminal health crises, my gratitude to the Lord has been fortified by the blessings I have received.

I have personally sinned and experienced tragedies, hardships, betrayals, and failures, yet our God is a good and loving father who is faithful and just. I have read His book, and I know the ending. That has given me great peace in life. When I was in combat, I never feared for my personal safety because if God took me, I would be in a better place. And what is peace? It is not the absence of war. It is the absence of fear. And when you have no fear, you have joy.

I am grateful to have not only the love of my immediate family members, but also friendships that have evolved into something even more special. Members of our extended family include an extraordinary couple, retired Army Lieutenant Colonel John (Jack) Czerwinski and his wife of over 57 years, Maggie. I met Jack when I returned from Vietnam and joined the Army Reserves 85th Division (Training) in Chicago. We were both Captains assigned to the Division Headquarters in the G-3, or Operations, section. Jack was a few years my senior and married with two young sons. He had met Maggie when they both were in the Army at Fitzsimons General Hospital in Colorado, when he was commanding a medical detachment as a Medical Services Corps officer, and she was an Army Nurse. Jack was from the north side of Chicago, where I had grown up, but we attended rival high schools. Jack went to an all-boys school called Lane Technical High School, and as I mentioned earlier, I went to the co-ed Carl Schurz High School. I could have gone to either, but Schurz was more of a college prep institution, and the opportunity to meet girls wasn't such a bad thing either.

Nonetheless, as adults, we quickly bonded and formed an amazing friendship. Maggie was from South Carolina, and there was always friendly banter back and forth about Yankee rudeness and Southern hospitality. Jack studied and majored in music, so he would often play the piano for us and sing with his beautiful tenor voice. He was a teacher in a boys' reform school in his suburb. They were devout Catholics and invited us to their home every year to share the Christmas holidays. Our younger children played with his boys, and our two families enmeshed.

The Czerwinskis became "Uncle Jack" and "Auntie Maggie." In fact, they agreed to be godparents for our children. If anything happened to us, we knew that Sumi and Jae would be in an environment of love grounded in faith and patriotism. Maggie and Jack moved to South Carolina when Jack retired as an educator and subsequently moved to the Phoenix, Arizona area where they reside today, but we remain in frequent contact. They are truly family members to us.

An individual in my life who exuded patriotism is the late Colonel David H. Hackworth, United States Army — Retired. I was honored to serve in the Army with Colonel Hackworth in garrison and combat, and we maintained a close friendship for over 40 years. He couldn't pronounce Mukoyama, so he called me "Mook," and I called him "Sir." His soldiers and friends affectionately knew him as "Hack." He was one of the most highly decorated combat soldiers in the history of the United States Army, having, among his numerous awards, two Distinguished Service Crosses (just below the Medal of Honor), ten Silver Stars; eight Bronze Stars; and eight (not a misprint) Purple Hearts. As previously menionted, he authored three best-selling books, *About Face*, *Hazardous Duty*, and *Steel My Soldiers' Hearts*. He was literally known as "Mr. Infantry" among the troops in Vietnam. Born on November 11, 1930, Hackworth was the youngest Colonel in the entire Army when he retired.

Hack chose to end his Army career so he could come out publicly to object to the continuation of waging a war that supported a corrupt regime at the cost of the lives of our nation's service members. He subsequently served as a military editor for *Newsweek* magazine and had a syndicated newspaper column. His example inspired me to testify before Congress. Even though that cost me my career, he taught me that leadership required making sacrifices and standing up for my people.

Hack contracted cancer due to Agent Orange and succumbed on May 4, 2005. I was honored to be the highest-ranking member in his honor guard burial contingent, which included Veterans who had served with Hack throughout his career. Before he died, Hack was one of the first reporters embedded with Army units in Desert Storm, and the troops adored him. Hack and his wife, Eilhys England, founded a 501(c)(3) non-profit called Soldiers for the Truth (SFTT) and maintained a website where service members of all branches could report on what was really happening and where the rubber met the road. In essence, he became the Ralph Nader of the military. As in combat, Hack fought for our military service members and nation to the very end. Born on Veterans Day, joining the Merchant Marine Service at the age of 14, fighting and bleeding in uniform, and continually demanding accountability of our leaders, Colonel David Hackworth epitomized every good and noble thing our flag represents.

These three examples of faith, family, and flag helped mold me as an individual.

We are in a state of spiritual warfare every moment of every day. There is indeed a guy named Satan, the Evil One, who is trying to wear us down in our faith to be eternally separated from a relationship with our heavenly father. My mantra is an arrow in our quiver in this spiritual battle.

So this is the source of my optimism. It's not difficult or complicated stuff. Remember, I am a simple-minded guy but grateful in all circumstances. Every day is a great day. I have my faith, family and live in the finest country in the world. My mantra gives me great joy when I say it because I see its positive effect on people and myself. Cling to it and share it as I do every day.

CHAPTER 21
AGENT ORANGE: THE POTTER'S GLAZING

The February 2012 morning of my heart attack was a life-changing point. It forced me to reflect upon my life and propelled me into the culmination of my life of service.

After the complications, operations, and necessary treatments, I began my recovery. I returned to work after a short time of recuperation at home and then went through a rigorous cardio-rehabilitation program. Upon completing the rehab, my kidneys' deterioration required that I undergo kidney dialysis, which involved going to a dialysis center three days a week to sit in a big reclining chair and hook up to a dialysis machine for 3 hours to purify my blood. I was on a list for donated kidneys but understood it could be years for a proper match. But after three months, I got a donor — cue another "God Thing".

The donor was our daughter, Sumi. She is not our biological daughter; both of our children were adopted. Yet she was a match. I did not ask my daughter to donate an organ — you can't ask your kid to do that — but she got tested on her own accord. My transplant operation was scheduled at the Northwestern Memorial Hospital world-renowned Kolver Transplant Clinic in downtown Chicago.

This operation placed an enormous amount of stress on my

wife. Much later, I learned that the day she drove home after my three operations, she stopped when she pulled into the garage and was overcome by emotions. She had a conversation with God, questioning why He would let this happen to such a strong man of faith, and she prayed fervently for my recovery. But, the stress became even greater on the day of the transplant operation. KJ had to watch our daughter being wheeled into the operating room and followed by me.

Sumi Mukoyama and John Ekholm wedding: Myself, KJ, Sumi, John Ekholm, and Jae

Our son was the glue that held everything together. Jae was in the health care industry, having his master's in nutrition and being a licensed Registered Dietician. He was our family medical "go-to" guy. But more than that, he was the man in the house during my absence, and KJ could depend on him in crises. Jae not only provided support through his presence to KJ during this tough time, but he was also there to comfort Sumi's husband, John. The three — Sumi,

John, and Jae — had a special close relationship, for which KJ and I have always been very grateful. They were all within 18 months of age, Sumi being the eldest. When Sumi and John were first married, they lived in Chicago and set aside one of their bedrooms for Jae whenever he visited the city. I called them the Three Caballeros (I was a Disney guy as a kid, who wasn't?). Jae hung out at the hospital, alternating between John and KJ throughout the day. He was the rock that provided them strength and comfort.

A short time thereafter, Jae told KJ that he was going to join the Army. He had tried before to sign up for Army ROTC when he entered the university. I had neither encouraged nor discouraged our children from joining the military. Of course, I would have been proud and supportive, but the decision was theirs to make. Jae was an excellent student, learned Korean Tae Kwon Do, and played ice hockey. He also played violin through high school. Upon meeting with the Army recruiter, he discovered he was ineligible for enlistment because, in middle school, he had taken Ritalin to treat ADHD. He was no longer taking it, but it was considered a disqualifier for Army service at the time. Jae was now in his early 30s, and the Army had removed Ritalin from their list of prohibited drugs used in the past. He was approaching the maximum age to join but could join the Medical Services Corps as a Registered Dietician. But KJ expressed that she needed him to stay and help her while I was recovering. Jae decided to cancel any further action, thus ending his chance of an Army career. He stood by and supported his mother in her time of need, and I was proud of his loyalty to his family over a potential military career.

KJ and I were so grateful that my illness experience was a family affair, and our loved ones rallied around us. Our son-in-law John had volunteered as a backup donor. During my immediate recovery period, our next-door neighbor, Jerry Menezes, drove me to the hospital for my treatments and appointments, with Bob Adams and my brother John also pitching in. KJ contemplated quitting her job to do those chores for me, but now she could continue working with seniors. Our neighbor, Jerry, insisted that KJ not quit her job, and that he would take care of me. Jerry and his wife, Luci, are immigrants from India and devout Catholics. They built their home shortly after we built ours in 1982 and have become dear friends. This is also true of our other next-door neighbors, Mary and Peter (Chris) Zouras, of Greek heritage. Our children went to school together, and like the Menezes, they have become like family over 40-plus years. They have always been there for us when we needed them and vice versa.

In our cul-de-sac, we all built our homes when the project was new, and our families all grew together. This burgeoning community reminded me of the neighborhood in Chicago where I grew up. As people of faith, we all knew and cared about each other in an integrated neighborhood of our Indian, Greek, Japanese, and Korean heritages — all fellow Americans — another true-life blessing.

Weeks before my transplant operation, my employer blindsided me with the news that after over 30 years together in three businesses, he no longer needed my services. Illinois is an at-will employment state, so the owner was within his legal rights. The severance agreement I had negotiated numerous years earlier wasn't a golden parachute, but it gave me sufficient resources to plan my next step in life. I had been a loyal employee with the respect of the company staff, but similar to my situation at the end of my military career, I did not desire to stay where I wasn't wanted nor appreciated, so I decided to move on. This was all part of God's plan, but I was too close to the forest to see the trees.

As I was going through this difficult valley in my life, I drew inspiration from the Job-like experience of a dear friend of mine, Milton (Milt) J. Blake, who was in the Army Reserves with me in the 85th Division when we were both majors at Division Headquarters. Milt, like me, was a Chicago native. He lived on the South Side, and his father-in-law was an influential pastor in the African American community. Milt and I became close friends with young families, doing that citizen-soldier juggling act of family, civilian job, and our military careers.

Milt went through a series of life tragedies that reminded me of Job in the Bible, except this was now. When I saw Milt at our monthly drill meetings, we shared what was happening in our lives. At one meeting, he mentioned that his father had passed away. At another meeting, I learned that his dear wife, Beverly, had a stroke and was hospitalized. Then he lost his job. It got to the point that I was reluctant to ask him how he was doing, but I needed to stand by him, listen, and pray for him. Then what would have been the coup de grace for 99% of the populace, the light of his life, Milton Blake Jr, his 13-year-old son, was diagnosed with sickle cell anemia. Milt Jr. was a huge baseball fan, and his favorite player was Reggie Jackson, so I bought a pictorial book about Jackson's career and visited with Milt Jr. at the hospital. He loved it. Not too many months later, Milt Jr. passed away.

Like Job, throughout these crises, Milt was steadfast in his faith and displayed a peace that carried him through. KJ and I went to Milt Jr.'s funeral on the South Side, and his African American faith community was absolutely strong in their support. KJ and I stood out in the crowd, and Milt and Beverly graciously greeted us. Milt was a rock of faith. I will always be grateful that the Lord put him into my life, so I could be faithful in God's plan for my life regardless of the circumstances.

In 2009, I became a member of a national Christian organization called Pinnacle Forum. Pinnacle Forum believes culture is forged in seven areas: religion, family, education, business, government, entertainment, and media. Highly successful Christian individuals in each sector would be invited to join Pinnacle Forum and form a network. We would influence our culture through our personal witness in the lives of our Christian values. The first step would be a personal transformation of each member, validating and growing their faith foundations through the study of the Bible. This initial stage of this process took about one year, to be followed by continuous study, much like continuing education in professional careers. The second step would be to utilize these educational and equipping skills to enable our members to affect cultural transformation.

After our local Pinnacle Forum group completed step one, we met and went around the table inquiring about each person's life activities. When they came around to me, I responded that I was active in my church, was helping my wife in graduate school, volunteering as an instructor of the Military Ministry of CRU (formerly Campus Crusade for Christ), leading a men's small group, and was co-directing a men's monthly church breakfast. But that was apparently not sufficiently busy. They asked me, "If you were King for a Day, what need would you address?"

My response was immediate. Our Military community—including Active Duty, Reserves, National Guard, Veterans of all eras, and their families—were hurting. We had been at war for over a decade, and our all-volunteer force comprised less than 1% of the United States population. As a result, our service members faced numerous combat deployments. The church needed to reach out to this community and offer appreciation, love, hope, and support.

We had failed to do that for my cohort, the Vietnam Veterans. But this vacuum was our chance to demonstrate God's love and healing. Immediately, four members raised their hands and said they were in. Military Outreach Greater Chicago was born.

Our concept behind Military Outreach was to develop a network of houses of worship of all faiths to support the members of their congregations who were on active duty, in the Reserve components, Veterans, or family members. A significant force in the formulation of our nonprofit was the person responsible for getting me involved in Pinnacle Forum in the first place. His name is Richard (Dick) Slayton, a highly successful founder of an executive search company from which he had retired. Dick came to the Lord well into his adult life and has been on fire to share his faith ever since by helping start up faith-based nonprofits. Dick was a member of our Pinnacle Forum group. He immediately recruited our Pinnacle partner volunteers and other strong candidates for a Board of Directors. He also advised on fund-raising efforts, personnel, and marketing. Military Outreach USA would not have been able to launch successfully without his mentorship, especially to me personally. Dick remains on the Board of Directors, and he and his wife, Donna, have been staunch supporters and encouragers to our team.

Howard Lang was another vital member of our original Board of Directors from Pinnacle Forum group. Howard and his wife, Mary, went to our church at the time. He is a retired Navy Captain with the Judge Advocate, or legal, branch of his service. We had attended Schurz High School in the city for a short time together, so our connection goes back well over 50 years, although we did not know each other at the time. Howard and Mary founded a legal practice that specializes in estate planning. Howard has served as Military Outreach USA's Secretary and legal advisor since its inception. He has served pro bono while devoting time and effort to participating in our activities. Another active program member was his wife, Mary, who has a spiritual gift for serving and a special heart for hospitality. KJ and I have been truly blessed to develop a close relationship with Howard and Mary. They are our trusted advisors regarding our personal estate, but more so, we are brothers and sisters in faith. They are treasured members of our extended family.

Based on my experience serving with the VA, one of the first questions I ask a Veteran is, "Are you registered with the VA?" Less than one-half of qualified Veterans are registered with the VA. Not having registered myself, I immediately corrected this oversight by registering at the James A. Lovell Federal Health Care Center in North Chicago. I had to apply the personal example principle of "do as I do, not do as I say." It proved fortuitous when I had my heart attack and kidney failure.

In working with Army Veterans and being one myself, I discovered that a large number of Vietnam Veterans had developed serious diseases associated with exposure to a chemical agent defoliant called Agent Orange. Agent Orange was a chemical herbicide used extensively in Vietnam to kill plants and foliage the enemy hid amongst to conduct ambushes. The VA had a list of presumptive conditions as a basis for VA disability claims. There have been 14 conditions identified, with research pending more. Two of those were Diabetes Type 2 and ischemic heart disease, two of my three conditions. My renal failure was a secondary effect caused by the heart attack.

In order to receive disability, I had to file a claim with the Veterans Benefit Administration (VBA). The kidney portion was initially denied, as it is not listed as one of the 14 presumptive conditions. I had to go through an appeal process which was later approved. However, there was an unintended benefit to me through this process. I could now better counsel Veterans about the claim process since I had gone through it myself. My street cred with Veterans was gaining momentum.

Military Outreach Greater Chicago grew to involve serving Veterans in numerous states, so we changed our name to Military Outreach USA. Our programs would be expanded into exceptional cross-country ventures. The final stage of the pottery had begun.

CHAPTER 22
THE INVISIBLE WOUNDS OF WAR:
TIMELESS LESSONS

Over the last decade, Military Outreach USA has matured and changed in its existence but maintained our core mission: demonstrating the love, compassion, and healing of Jesus to the military community. With these objectives in mind, we have found ways to support Veterans in two of the most egregious challenges: homelessness and suicide.

There are thousands of well-meaning nonprofits to serve the needs of our military. Our Board of Directors has always understood that we could not provide all the necessities to all Veterans in need, so we were determined to let God guide us by providing opportunities. Consequently, our two core programs would address the gaps in the system. Our first program aided homeless Veterans, specifically, those coming out of homelessness, through the aid of social workers of the Department of Veterans Affairs. *The Veterans Exiting Homelessness Program*, or VEHP, serves one of our most necessitous populations. Essentially, when a social worker feels a homeless Veteran has been stabilized and is ready for permanent housing, they are given the keys to a one-bedroom apartment subsidized by a HUD-VASH (Housing & Urban Development Veterans Assistance

Special Housing) voucher. But the keys are all they get. Imagine the first time you moved into an apartment — you *need* stuff. But the VA is not funded for that purpose. We saw that gap and decided to take action. We reached out to our network — houses of worship, Veteran Service Organizations (such as the VFW, American Legion, and Vietnam Veterans of America), philanthropic groups (such as the Rotary, Lion Clubs, and Rolling Thunder) and high schools to donate new items from a list we provide called *Move-In Essential,* or MIE, items. In 2016, the Department of Veterans Affairs signed a Memorandum of Agreement with Military Outreach USA to establish this program.

The MIE collections are supervised by one of our most dedicated volunteers, Jim Dunne, a Vietnam-Era Veteran with a great team of volunteer Veterans actively working with churches, schools, and VSOs. They work in conjunction with a UNITS Moving and Storage facility in Mundelein, Illinois, owned by a patriotic citizen, Tom Koenig, who has generously provided the use of his storage facilities, to include pick-up from collection sites throughout the area which our supported VA hospitals then pick up from his facility. All of this is donated by the volunteers and Mr. Koenig.

Another part of the VEHP program is called *Beds for Vets.* These homeless Veterans do not have beds. They might have a sleeping bag, poncho, blanket, or cardboard box but no bed. So, we ask our network members who cannot conduct a collection of MIE items to consider making a tax-deductible donation to Military Outreach USA. These funds are specifically designated for providing beds, pillows, sheets, and pillowcases. The Beds for Vets program is managed by our Executive Assistant Sue Brown, who, with her husband Mike, devotes time and talents far beyond the minor part-time consultant fees she receives.

How have we done? As of this book's writing and since the program's inception, we have collected MIE items valued at over $1,300,000, distributed to over 55,000 Veterans, and delivered over 2,600 beds to VA facilities nationwide. All items are provided free to the Veterans and their family members.

Our second major program deals with the unacceptably high rate of suicides among our Veterans and active-duty and Reserve component service members due to the invisible wounds of war. Three of these wounds are well known to the general public — Post Traumatic Stress Disorder (PTSD), Traumatic Brain Injury (TBI), and Military Sexual Trauma (MST). Less publicized but most insidious is a condition called Moral Injury (MI). The concept of Moral Injury is so intuitive that you will grasp it in 30 seconds.

From birth to adulthood, you develop a personal moral code, a sense of right and wrong. That could come from your faith, your family, or your community. Then, you may join the military and learn a warrior code. As part of your duty and service, you might have to participate in actions or activities that violate your personal moral code, such as killing another human being. At that time, you sustain an invisible wound of war — moral injury. It's not a physical wound. You can't see it. (See, that was less than 30 seconds.) But with military operations, you are constantly moving from point A to point B to point C. You don't have time to stop and reflect on this stuff. So what do you do? You bury it. It can lead to unresolved guilt, shame, depression and suicide.

Moral injury is nothing new. In the Bible, the Book of Numbers, Chapter 31, you will read how the Israeli warriors returned after defeating the Mideonites. But Moses would not let them reenter the camp without going through a purification process. The knights returning from the Crusades were not permitted to participate in the Holy Sacraments until they went through repentance and reconciliation. The Native American Indian culture has had community sweat lodges to cleanse before and after a battle. Society has known for millennia that if you send warriors out to war, you must help them reintegrate upon their return. Yet our modern society has forgotten all of this obligation.

At Military Outreach USA, we believe moral injury is addressed with the forgiveness and grace of a moral authority, a loving God, the counseling of clergy and/or sensitive therapists, and the support of a community offering hope and help, rather than by a medical doctor with prescription drugs.

In 2015, Military Outreach published a book all about moral injury titled *They Don't Receive Purple Hearts*. Our former Executive Director wrote it, and I was the editor. Neither of us was a social worker, doctor, or psychologist; we were just two Vietnam combat Veterans who had walked in the same boots as those experiencing moral injury. Through easy-to-understand language and simple concepts, the book intends to educate and inform Veterans, their families, mental health professionals, and the public on this crucial subject. If you go to our website at www.militaryoutreachusa.org, you will find a .pdf file of the entire book you can download for free. Military Outreach USA does not charge for our services; we only intend to serve those who have served.

As we wrote in the book, "There are no medals or ribbons for those who suffer from the invisible wound of war. There is only the torment and struggle carried within the darkness of the soul. It is a house of worship through which one who suffers can begin to see the light."

That same year, I viewed a copy of a DVD titled *Invisible Scars* produced by another faith-based organization serving our military called Crosswinds Foundation based in Birmingham, Alabama. The subject was PTSD, and it was the most impactful educational film I had seen on the subject. I immediately contacted Crosswinds Foundation to congratulate them on their film, which they provide free of charge to any military community member, including family, who request a copy. During that conversation with the founder, Bob Waldrep, who has a heart of gold for our military and whose father was a Veteran, I asked if they were planning to do a DVD on moral injury. Bob was familiar with moral injury but explained that they were, like Military Outreach USA, a small non-profit with limited resources. However, I countered and reminded him that with a suicide rate of 22 plus Veterans alone per day, this issue required immediate action. Such a film would be critical in providing hope to our Veterans and their families. Bob said he would bring it up to his board for consideration.

Weeks later, I received a call from Bob saying they were moving forward with a new DVD about moral injury called *Honoring the Code*. He then asked me if I knew any Veterans who would be willing to have their stories filmed about moral injury. I knew a ton of Veterans from our outreach programs at the three VA Medical Centers in the Chicagoland area, so I was able to find numerous Veterans from all the services, both male and female, to participate in the filming.

The Crosswinds Foundation production crew came to Chicago, and they conducted filming. As with "Invisible Scars," this DVD included interviews of actual Veterans giving their names, services, and wars. Some spouses are also interviewed. There are social workers, psychologists, and military chaplains included. If you watch the second DVD closely, you might see someone you recognize.

This meeting of our two organizations was another "God

thing." What were the odds of an Alabama-based group and a Chicago-based group unfamiliar with each other having identical missions and combining efforts to bring hope and healing through their mutual faith foundations? Only God. And Military Outreach USA also sponsored a panel on moral injury that appeared on the Pritzker Military Museum and Library *Citizen Soldiers* TV Series that appeared on Public Television.

In 2020, amidst the COVID upheaval, Military Outreach USA expanded to include a new outreach program, Stand-By-Me Heroes (SBMH). This organization was an already existing nonprofit organization founded by Philip "Chappy" Ferrer, a Desert Storm Army Combat Veteran who is an ordained minister and a chaplain for several Veteran organizations. SBMH includes Veterans called Foxhole Soul Counselors (FSC). They are not licensed counselors but fellow Veterans or Veteran family members who walk alongside those dealing with the invisible wounds of war. Their offices are jokingly referred to as Dunkin' Donuts. They listen, develop a trustworthy relationship, and offer hope and help with the appropriate Veteran services if requested.

My introduction to Chappy was another "God Thing." We had recently lost our previous Executive Director and desperately needed a replacement. I was at my automobile dealer and, as I do every day, was wearing my Vietnam Veterans cap. I do this for two reasons: (1) To let people know that we have Veterans everywhere in our communities, and (2) It invariably brings up conversations with other Veterans. That day there was another Veteran who was also wearing a Vietnam Veterans cap, so we introduced ourselves. I spoke about Military Outreach USA and our mission. The Veteran responded that he knew a fellow American Legion member doing exactly what we were doing. He gave me Chappy's contact information, and I called him immediately. The rest is history, and Chappy now serves as our Executive Director.

Crosswinds Foundation has developed a training program for organizations to equip and enable their members to help Veterans understand and deal with their invisible wounds of war. It is called Centers of Hope, and their work manuals provide Military Outreach USA with materials for local community use. Crosswinds Foundation also has published a book entitled *Hope for the Warrior Family*, addressing the circumstances and needs of the military family. Military Outreach is assisting in filming a third DVD accompanying that book, filling another gap in our program offerings. It provides an opportunity to develop military community "small groups" that can continue the healing and hope after completing the Centers of Hope materials.

Military Outreach has further participated in several healing retreats for our wounded warriors and family members. These programs range from a one-day event conducted at the Bellarmine Retreat Center in Barrington, Illinois, a Jesuit Retreat Center (attendees do not have to be Catholic) to a four-day retreat at the Union League Club in downtown Chicago, to another four-day retreat in northern California.

I got involved in the first retreat when a Military Outreach USA Director, father Matt Foley, invited me to join a committee to create the retreat, especially for Veterans, in 2016. Yes, his familiar name was used in Saturday Night Live by the late Chris Farley, who was a high school classmate of Father Matt, who was an Army Chaplain who was deployed four times to the Middle East with the 101st Airborne and Special Operations Forces. He joined the Army when his brother, a career Army officer, was deployed. Serving in his 50s, Fr. Matt completed paratrooper training or Jump School. We

205

have conducted the Coming Home Retreats thrice a year ever since, only with an interruption due to COVID restrictions in 2020. It is a spiritual healing experience that does not emphasize any specific religion and is therefore open to all comers. It is for all Veterans and their loved ones. It has been life-changing for many participants and life-enriching for all, including the staff.

Another former Military Outreach USA board member, Victoria "Vic" Bruner, RN, LCSW, invited me to participate in an annual four-day retreat experience called *Bridging the Gap* funded by the Walter Reed Foundation at the Union League Club of Chicago (ULCC). It was a prestigious private club with a history of patriotism and community involvement since its foundation by members of the Grand Army of the Republic, the first national Veterans organization after the Civil War. The ULCC members, led by the Club's American Legion Post, generously provided accommodations for the retreat staff and participants, including their five-star cuisine. Vic was the widow of a pilot who was killed in action in Vietnam, and she has devoted her entire adult life to serving our military. She was at the Pentagon on 9/11 and rendered medical and psychological aid. Over the years, I have been honored to be designated as an "Elder Warrior" at the retreats.

The third retreat modality I participated in, again as an Elder Warrior, was the Healing Journey experience conducted by the Healing Through Reckoning and Responsibility organization, located in San Luis Obispo, California. The retreat is led by father Michael Cicinato, a retired parish priest, and Peter Steinberg, LCSW and P.C. The retreat is conducted at the St. Francis Retreat Center in San Juan Bautista, California. Like Bellarmine, it is a spiritual healing experience incorporating many ancient Native American traditions. In an incredibly moving ceremony, non-Veteran participants act as representatives of our society and take responsibility for sending our warriors into combat and welcoming them home.

These are just a few examples of healing opportunities available to our Veterans and their families. Quantifying the exact results of these efforts is nearly impossible due to confidentiality and privacy laws. In order to gain the trust of those being served, we often do not get their names but rather communicate by using call signs. But countless Veterans and family members have been aided, specifically to the point of preventing planned suicides. Recently, we met two Marine Veterans who had set dates for their suicides with weapons and their dress uniforms pressed and ready to wear. But through interaction with our FSCs, they were given hope to persevere. Some have experienced divorces and lost children, but they have recovered to the point that they are now Foxhole Soul Counselors. I have had Veterans personally thank me for saving their lives. We know we are making a difference. Frankly, our efforts are worthwhile if we make the slightest dent in reducing the rate of Veteran suicide. Every life saved continues to fuel us now and in the future.

Individually, I travel throughout the country making presentations about moral injury at VA medical centers, churches, medical colleges, and active-duty installations. I am busier in so-called retirement than I was when I was working full-time. And I am happier. How can that be? For the answer, check out the next chapter.

CHAPTER 23
LIVING ONE'S CALLING: IN THE SWEET SPOT

God, the Master Potter, fashions each of His children into a unique creation with individually unique spiritual gifts that lead to our callings. After 67 years of experience and qualifications, I was ready — molded and shaped to be the vessel designed to fulfill my life's mission.

Before starting Military Outreach USA, I had served in many different roles and carried substantial responsibilities. I was obtaining varied experiences that would all be brought together for one purpose, to reflect God to the world, specifically to people of faith, families, minorities, and the military.

When you are living your calling, people notice a certain aura of contentment and deep passion about you. This satisfaction doesn't mean that your life is always full of milk and honey, as we are constantly battling the evils of pride and temptation. But when people are clearly focused on their life pursuits, such individuals emote something special in their mannerisms, treatment of fellow human beings, and the daily conduct of their lives. There have been so many in my life, but I will give examples of three individuals I have seen live God's calling.

In all of the examples, the individuals had initial careers that provided them with experiences and resources to pursue their final paths. The first individual is a fellow Vietnam Veteran, John Domagata, or "Bud" as we call him. I have known Bud and his wife, Holly, since my Willow Creek days. Bud and I worked for years together in men's ministry activities, having been in my men's small group. He also was a strong encourager and participant in the military ministry effort at Willow and the early days of Military Outreach.

Between our shared combat experience and desire to help Veterans, we had created a special bond and considered ourselves brothers in faith and arms. We would literally die for each other. Bud's childhood was the origin of his life's calling. Growing up, he spent every summer on the banks of the Wisconsin River, where his grandfather had built a cottage. His father-in-law owned a sporting goods business where Bud worked for over 20 years. Bud loved powerboating and water-skiing on the river, which for all intents was like a lake to me. After becoming a successful long-term care insurance salesman, he purchased his grandfather's cottage.

When Bud approached retirement age, he and Holly tore down the old cottage and built a beautiful home on the site. Over Bud's objections, I refer to it as the Castle on the Lake. Bud and Holly then combined their love for God and the outdoors to create a water-skiing ministry. They initially targeted younger skiers and recruited them to join on Sunday mornings prior to church. Bud and Holly's home has become a beacon for young people, and throughout the many years of the ministry, many have come to faith through their efforts. They have also been pillars of their church, which had just started when they moved to Wisconsin and is now a thriving spiritual community. Bud has continued to be very active in men's ministry and Veterans' activities — like someone else you may know — but he has found his true calling by combining his life-long passion for the water and faith. Whenever we talk, it brings a special joy to my day.

The second person is much younger and could almost be my son. His name is Justin Kron. He is a Messianic Jew — a person of Jewish heritage who has accepted Jesus as his Savior and Lord. He and I were in a former Pinnacle Forum group together, and we also attended the same church in our community of Glenview. The northern suburbs of Chicago near Lake Michigan have a sizable Jewish population. Justin and his wife, Judy, have spent years in outreach to the Jewish Community through an organization they founded called the Kesher Project. This organization holds a monthly gathering to present programs involving subjects relating to the Bible and its Jewish roots. The program is open to all in the area who are interested in Judeo-Christian connections. For years, Justin has had a heart for reaching out to the Jewish people in Israel and recently released a movie, *Hope in the Holy Land*, about the Israeli-Palestinian situation. It has received very positive reviews and has won a prestigious film award. Religion has always been at the core of Justin's life, but if you met and interacted with him, you would see that his current pursuits are evidently his true purpose.

As you may have anticipated, my wife, KJ, is my third example of someone living their calling. Her spiritual gift of serving has been evident throughout our marriage. Our daughter Sumi has always marveled at how total strangers would walk up to KJ in a crowded restaurant to ask for advice or help. This miracle happened more than once because KJ has an unmistakable aura of compassion and goodness. But, like me, God was working on preparing her for her true mission. Her experience as the caregiver for my parents, her studies of Gerontology for her master's degree, and our hospice volunteer experience laid the foundation for her serving the elderly. When most people were retiring, KJ started her 16-year career in a rehabilitation center for elderly residents. Even today, she continues her calling by caring for a 94-year-old Japanese widow. All three of these people have a common foundation of their faith, and their unique spiritual gifts have empowered them to live their callings. I am in awe whenever I interact with these individuals and witness them in their pursuits. Their entire lives have culminated in bringing them to their callings. They are truly thriving in their distinctive service endeavors.

I, likewise, feel so blessed as I enter the fourth quarter of life. And the Good Lord has continued to provide additional significant serving opportunities that advantage my life experiences.

One involves a dear friend, Larry Spatz, who contacted me out of the blue — another God Thing — when he saw our website on the internet. Although he is not a Veteran, Larry has a heart for our military. One night, he saw a program on public television about Gold Star families, and he created a nonprofit with a line of patriotic clothing to benefit the children of fallen heroes. Larry, who is Jewish, lived in the North Shore suburbs, and we immediately bonded because we both grew up in Chicago and were virtually the same age. My knowledge and background of the Jewish traditions and community also struck a familiar chord. Larry had created a nonprofit with a disabled Veteran called Enabled Enterprises. Its mission was to develop entrepreneurial opportunities for Veterans, especially the disabled, who could formulate local branches selling and servicing products developed by Enabled Enterprises. One of the products, electric bicycles for police departments, has been marketed and sold to some departments in our area. They also designed a motorized platform for wheelchairs that permits those in wheelchairs to go on rugged terrains and beaches. After Enabled Enterprises had been in business a short time, one of the co-founders, a Desert Storm Marine Veteran who lost both legs, tragically died by suicide. This tragedy spurred Larry to dig deeper into Veterans' issues and see what he could do to help. After learning about moral injury, Larry was captivated by this explanation and connected with Military Outreach USA to support us. I continue to work with Larry in his efforts. As a result, we have become very close friends, as well as his wife, Marilee, with my wife KJ.

Another serving opportunity arose when I was approached by Ken Clarke, then the President of the Pritzker Military Museum & Library (PMML), regarding their Board of Directors. The PMML is a hidden treasure of military history and artifacts in Chicago. It is located on the corner of Michigan Avenue and Monroe Street, across from the Art Institute of Chicago and Millennium Park. It was founded by Col Jennifer Pritzker, who retired from the Illinois Army National Guard. Col Pritzker had served on active duty with the 82nd Airborne. I had been brought to her attention by John Schwan, a close friend of Al Lynch (the Medal of Honor Recipient who was on the original Board of Military Outreach USA), who, like John, served in the 1st Cavalry Division in Vietnam.

After I met John on several occasions, we became close friends through our similar histories and values. John is the founder of CTI Industries Corporation, a manufacturer of zip lock bags and helium party balloons listed on the NASDAQ exchange. John and his wife, Beverly, are a wonderful couple and have been generous benefactors, especially to organizations serving our Veterans and our Nation, including Military Outreach USA and Enabled Enterprises. John and I are similar in age, he being slightly longer in tooth than I, and we both grew up on the North Side of Chicago and attended rival high schools. He is also a Chicago Cubs fan, while I am of the Chicago White Sox persuasion. Yet, we have become like brothers through our shared experiences in the Army in Vietnam and, most importantly, our shared values of faith and family.

I was honored to be invited to join the PMML Board and have been a member since 2015. In addition to a world-famous military literature writing program, the PMML has an extensive series on military subjects on public television. They are even undertaking a major project to build a state-of-the-art archive center and a National Cold War Memorial near Kenosha, Wisconsin. As my association with Colonel Pritzker has deepened, my admiration and appreciation for her philanthropic endeavors have grown exponentially. She has personally supported our nation's youth by encouraging the JROTC program in Chicago schools, which has inspired countless students from the inner city to not only graduate but excel academically and contribute to their communities. Jennifer virtually single-handedly was the driving force for this nation's Centennial observance of the end of World War I and the creation of the World War I Memorial in Washington, D.C. And her leadership and generosity are also evident in her involvement in the Elizabeth Dole Foundation, which supports military family caregivers, and her contributions to the recently opened Museum of the United States Army in Virginia. She is a true patriot, and I am honored to serve with her at the PMML.

Finally, two other of my serving activities involve medical service to our Military Community. Veterans' health should be assured and prioritized because Veterans have already sacrificed so much for this country. Six years ago, I was asked to co-chair a Patient Advisory Council (PAC) (since renamed the Stakeholders Advisory Council [SAC]) of the James A. Lovell Federal Health Care Center, or FHCC, in North Chicago, Illinois along with Al Lynch. As you can see, I have benefited significantly from Al's coattails. The Lovell FHCC is the nation's only combination Department of Defense (DOD) and Department of Veterans Affairs (VA) hospital. The Council sought input from non-military volunteer organizations supporting the FHCC regarding the facility's operations. From our first meeting, I made it clear that our mission was to improve the experiences of the patients at Lovell to the best they could be. I expected recommended solutions for any problems, not just complaints. The hospital Director (a VA senior medical executive) and the Deputy Director and Commanding Officer of the hospital (a senior Navy Medical officer) attend the meetings and have been fully supportive.

The current hospital Director, Dr. Robert Buckley, is a retired Navy Captain and M.D. who was the Lovell FHCC Deputy Director and Commanding Officer of the hospital when the then Patient Advisory Council was formed. He then retired, worked at another VA medical center, and was appointed the Lovell Director. He was the perfect choice based on his knowledge and experience, and he earned confidence in the facility and staff. I joke with him that out of the VA system of 130 hospitals nationwide, he is the only hospital Director with an address on the street with the Director's name. The Lovell FHCC is on Buckley Road. This serendipitous name further proves that the hospital is in the right hands.

The other serving opportunity is my appointment to the Board of Directors of the Friends of Fisher House of Illinois. There are over 90 Fisher Houses nationwide co-located with DOD and VA Hospitals. Their function is synonymous with the Ronald McDonald Houses, except this is for our military and Veterans. It is a nonprofit organization that supports the Fisher House Foundation, which pays for all the construction costs of the houses. For patients at Lovell undergoing serious medical treatments living more than 40 miles away from the facility, their families are provided living quarters at no cost to them. We will be building the Lovell Fisher House next to the Lovell FHCC, and we are raising the funds with ground-breaking scheduled soon.

Speaking of Lovell regarding both the FHCC and the Friends of Fisher House, another blessing in the final quarter of my life has been having the privilege of being introduced to and working with Capt. James A. Lovell, U.S. Navy – Retired, an actual hero and Apollo 13 astronaut. A true gentleman, he lives in the northern area of Illinois close to the Naval Station Great Lakes. He has generously donated his support to the Federal Health Care Center named in his honor and is also a Friends of Fisher House of Illinois Director. He is a strong man of faith and continues to demonstrate that both in his church and community. As he has with millions of our citizens, he has inspired me to carry the torch of American values forward.

Together with my ongoing Military Outreach USA, and church activities — especially my men's groups — my serving opportunities cup runneth over. Nothing gives me greater joy. When I thought I knew the answer to prayer in college regarding the rifle or the cross, the answer wasn't no. God was telling me to wait. My wait is over, and I am in my sweet spot.

CHAPTER 24
OLD GLORY: BEACON, UNIFIER, INSPIRATION

Throughout history, people have had symbols that have caused them to voluntarily devote their loyalty, even their lives, when called upon. Jewish people have the Star of David, Christians have the Cross, and warriors look to the flag in battle. We Americans have "Old Glory."

The flag of the United States of America, "Old Glory," also known as "The Star Spangled Banner," is much more than a piece of cloth. Since civics is no longer a required subject for graduation in many of our schools, I am duty-bound to address this subject for present and future citizens of our nation.

Old Glory is a symbol of the greatest nation in the history of the world. The United States' most sacred treasure is the lives of our service members, and we have sacrificed more of our own than any other nation fighting to defeat evil in this world on foreign lands on behalf of other nations and people. Especially in the Twentieth Century, we fought to defeat oppressive totalitarian forces driven to conquer others — from German, Italian, and Japanese fascism to North Korean, Chinese, and North Vietnamese communism. Had we not prevailed in World War II, much of the world would be speaking German and Japanese today. In all these cases, we did not fight to conquer their territory or the people in those lands. In fact, the only land we received was for cemeteries to bury our fallen warriors. The United States also sacrificed over one-half million soldiers fighting the Civil War to eliminate slavery and maintain the nation as one United States of America. The flag, therefore, represents freedom for all people. That is why oppressed people throughout the world often display the Stars and Stripes in their battles against oppression in their own countries. We remain the beacon of freedom across the globe.

We also are the beacon of hope. Whenever there is a natural tragedy or event in the world, both the citizens and government of the United States step in to provide humanitarian support. From rebuilding post-World War II Europe through the Marshall Plan to aiding Haiti after its disastrous 2010 earthquake, the United States has provided food, money, people, and supplies to those suffering. Many of these efforts are from our faith-based organizations and individual houses of worship. Though often overlooked, the United States military assists medical, emergency, and logistical needs during crises. Our humanitarian giving is in the basic DNA of Old Glory. It is not for profit or power motives — strictly charity based on our values to help others. As a nation, we recognize that of those blessed with much, much is expected to help others.

Through the genius of our Founding Fathers, the United States is also the most successful experiment in nation-building. They crafted the Declaration of Independence and the Constitution of the United States, the oldest constitution still in effect today that remains the gold standard. By recognizing the inalienable rights bestowed upon all people by our Creator — through the Declaration of Independence, the Constitution Articles, the Bill of Rights, the First Ten Amendments of the Constitution, and subsequent amendments

— the concept of freedom and equal opportunity is embedded in the fabric of Old Glory.

The Founders recognized that for the government to succeed, the power had to reside in the people, not the government. They had experienced the oppression of a government empowered to suppress its citizenry. Yet they were wary about ruling by the whims of trending public opinion in a pure democracy, where a slim majority can crush the minority. Accordingly, the Constitution describes a representative form of government, a republic with elected representatives controlled by a system of checks and balances through the three separate but equal branches of government — the Legislative, the Executive, and the Judicial branches.

Is this setup perfect? No, but the Founders provided for corrections with the amendment process. Again, these amendments require careful discussion and study involving all three branches of government to protect against a quick decision based on a popular whim.

The origins of Old Glory go back to the Revolutionary War. A symbol worthy of representing the new nation and its values would rally its people and service members alike. It instilled a sense of pride and unity. General Washington himself was involved in the final design. They decided upon a blue field with the number of stars representing the then-13 states (providing for additional states in the future) and 13 alternating red and white stripes representing the original thirteen colonies. The red color represented valor, the white represented purity, and the blue represented justice. The "Red, White, and Blue" has inspired generations of Americans to the concepts valued by its citizens. At one time, the flag was flown in battle at the front of the column of a military formation so that the soldiers could follow and see it as a rallying point. The term "red, white, and blue" has become synonymous with Old Glory and, therefore, the United States rallying cry has been a call to the colors referring to answering our nation's call to duty. As those of us who have worn our nation's uniform often say, "These colors never run."

This ethos used to be reinforced daily in our schools by reciting the Pledge of Allegiance. I recall fond memories of standing proudly at my desk at the beginning of each school day with my classmates, standing straight, putting my hand over my heart, and reciting the Pledge. It gave me a sense of pride to be a part of something bigger than myself, an American with all my classmates of all races, religions, and ethnicities. I recommend that parents should check with their schools, encourage them to say the pledge daily, and voice support for this time-honored tradition.

Every day – except for in inclement weather – I fly Old Glory in front of our home. I respectfully salute the flag each time I raise and lower it, but in a slower motion at the end of the day. I try to mimic the ritual on all active military bases, where there is a daily flag-raising and lowering ceremony. The flag-raising is early in the morning at 0600 hours (6:00 AM) with the bugle sounding Reveille. In the evening, at 1700 hours (5:00 PM), there is a "retreat," when a cannon is fired, and the flag is lowered slowly with the playing of the National Anthem. When the cannon sounds, everyone outside stops their activities. All cars on the post stop, and everyone gets out of their vehicles. All face the flag pole, stand at attention, and salute if in uniform or place their hand over their heart if in civilian clothes.

The history of our national anthem goes hand in hand with the symbol of Old Glory. The words of *The Star-Spangled Banner* were written by Francis Scott Key during the War of 1812, where he witnessed the 25-hour unrelenting bombardment of Fort McHenry in Baltimore Harbor. The British had demanded that the Americans lower Old Glory to signify their surrender or the Fort would be reduced to rubble by the British armada assembled outside of the harbor. Key was confined on one of the British ships, negotiating an exchange of American prisoners of war when the British issued their ultimatum. He relayed the news to the prisoners in the hold of the ship, and they prayed for the soldiers at Fort McHenry and their families. Throughout the night, Key observed the horrific sight of the Fort being hit time after time. Although badly ripped and torn from cannon fire, the American flag remained flying. When the sun rose at the end of the bombardment, the flag was still there, flying high. The Americans on the ship cheered and saw it as a sign that their nation would prevail. The flagpole did indeed take numerous direct hits, but soldiers kept the flagpole up, at the cost of some of their lives.

In the final analysis, in war, except for the Japanese in World War II, surrender occurs when one soldier physically occupies the land of another soldier. Without a strong army, the British withdrew. The Infantry, the Queen of Battle, reigns (spoken like the true Infantryman I am). The military has two strong traditions that instill the concepts of courage and selfless service. First is the professional military associations of the three main U.S. Army combat arms — Infantry, Armor, and Artillery. Each has an association by which its soldiers aspire to be recognized. For the Infantry, it is the Order of Saint Maurice; for the Armor, it is the Order of Saint George; and for the Artillery, it is the Order of Saint Barbara. Saint Maurice was an Egyptian Christian in the 3rd Century A.D. who commanded a Roman Legion in Thebes. When the Roman emperor ordered the persecution of Christians, Maurice and his men refused, and all were executed. At their execution, they famously outcried, "It is better to die innocent than to live guilty!" I have the unique distinction of having been elected to all three orders.

The second tradition is our military challenge coins. These coins are a relatively recent popular development, having begun less than one hundred years ago. It originated in elite Army units, such as the Rangers and Special Forces, and today is universally practiced in all military service units, as well as first responders, police, and fire departments. The coins represent the particular military unit and are awarded by the unit's commander or senior enlisted soldier. It is an honor to receive the coin; the tradition is that you must always carry it with you. If you are at a restaurant or bar, and another member of your unit pulls out their coin and places it on the bar, you have to place your coin on the bar too. Failure to produce your coin results in you buying the challenger a drink. But more importantly, it sustains the special bond of community and fellowship associated with the unit. You don't buy these coins but are given them as a symbol. My 32-year career of extensive active and Reserve component assignments has resulted in my being given coins from unit leaders and members throughout the Armed Forces. Nowadays, my collection is unique and quite large. The first two pages of the list of my growing collection of over 350 coins is attached as Appendix J[12].

There are countless other stories and examples of courage by America's service members, with the flag as their inspiration. The famous painting of General Washington crossing the Potomac River at Valley Forge with the *Stars and Stripes* alongside him in the boat comes to mind. In our family room, we have hung a Civil War battle painting that depicts the charge of the 1st Irish Brigade of the Union Army at Antietam, led by the commander on horseback alongside

---

12 Examples from my Challenge Coin Collection are found in Appendix J

Old Glory carried by flag bearers entitled, *Raise the Colors and Follow Me!* I also think of the iconic photos of the Marines in World War II raising Old Glory on Mount Suribachi at Iwo Jima and the flag planted on the Moon by our brave astronauts. These moments of captured history evoke feelings of pride in the sacrifice and dedication of those who have worn the uniform of the United States of America.

And patriotism involves more than serving in and supporting our military. You can serve in other ways, so long as it honors the true spirit and ideals for which our nation stands. During the Vietnam War, there were many protesters of my generation. Some were sincere conscientious objectors, but the vast majority were trying to avoid the risk of being wounded or dying and therefore ran away to Canada as draft evaders.

But other patriotic Americans objected in principle to the war itself but decided to honorably serve our country in alternate ways. My cousin, Wesley Mukoyama, instead volunteered for the Peace Corps and served a tour in Africa in Tanzania at the same time I was serving in Vietnam. We used to send letters of encouragement to each other. We both admired each other for serving our country and following our principles. We were proud of one another. Wesley, a social worker, and his wife, Janice, continued to serve in the Peace Corps for years, and he later served at the VA Medical Center in Palo Alto, California.

Patriotism is honoring our nation represented by the *Stars and Stripes.* It is therefore not blind jingoism, or connected to any particular political party, but rather reflects this great country's principles, values, and accomplishments. It has nothing to do with war-mongering or imperialism. It honors the sacrifices and successes that made this country the greatest on earth.

This is why I place Old Glory on a flagpole in the front of our house every day and respectfully bring it in at night, wearing my Veterans cap and respectfully saluting each time. That is why the flag is displayed prominently in parades for our national holidays on Memorial Day, the Fourth of July, and Veterans Day. That is why when the flag passes by during parades, Veterans in wheelchairs will rise, if possible, and all will salute to honor not only the flag but the brave women and men who have defended and are defending what it represents.

Loyalty to Old Glory demands that we take individual responsibility as citizens and be eternally grateful for the rights and privileges, the freedom and opportunity that it symbolizes so that we can remain "the land of the free and the home of the brave."

CHAPTER 25
JOY: THE SOURCE

People have often asked me, "Jim, why are you so happy? How are you so optimistic and peaceful? What makes you see the glass half-full instead of half-empty?"

Peace is not the absence of war but rather the absence of fear. And when your life is not controlled by fear, you have joy. I have learned and stood by these two definitions through my faith, and they give me a steady peace of mind. This mindset doesn't mean that I am always wearing rose-colored glasses, nor is it the case that I have never doubted my abilities or the outcome of a critical situation. Doubt is not bad. It generates a realistic assessment and plan of action. But I understand that God has a plan, and I do not fear what he has planned for me.

This book has no acknowledgments section. This decision was made with the intent that the entire book would reflect on my life's accomplishments due to the incredible people that God placed in my journey. For those that I was not able to acknowledge by name in these pages, please understand that just because you are not explicitly mentioned, you are neither forgotten nor unappreciated, especially those in my faith-based small groups; those walking with me in our non-profit serving organizations; and my brothers and sisters who wore our nation's uniform with me in peace and war.

I am eternally grateful to everyone who has contributed to my maturing as a man of faith; participated as members of my extended family and friends; and, most importantly, my God. Throughout my life, there have been so many inspirational examples of grateful people.

In 2018, I attended a conference in San Diego to represent Military Outreach USA and had the honor of hearing and meeting one of my life's most inspirational speakers. Her name was Dr. Edith Eva Eger, a survivor of the Nazi death camps. In her 90s, she has a persona of elegance and joy when she enters a room. As a teenager, she survived the death camps and brutal, deadly marches aided by her fellow prisoners who saved her life. After the war, she earned her doctorate in Psychology and became a renowned motivational speaker. She even is a training consultant to the Navy and Army regarding PTSD.

But most importantly, she is the epitome of gratitude. She has written a wonderful book entitled, *The Choice: Embrace the Possible.* Dr. Eger embraced the choice to forgive and move on with life with a positive attitude. If a person who survived the worst of humanity can do that, it is possible for all of us.

Although I only met Dr. Eger in my later years, I am proud to have lived with a similar attitude. In my lifetime, I have faced obstacles that I could have used as excuses. I could have dwelled on my short stature, minority status, or lower-class background, casting myself as a victim. But if I had spent all my time contemplating these disadvantages, I would never have had my success or blessings. I chose to be optimistic about my future and remain grateful for the past and present.

Life is a constant experience of choices. It is crucial to understand that you alone control your choices and how you react to your circumstances. A major advantage falls to those who appreciate the blessings they have received in life rather than those who accept the "woe is me" attitude and blame their unfortunate circumstances on others.

With Dr. Edith Eger. San Diego, CA. 2018

Especially in America, the victimhood explanation doesn't fly. There are numerous government and non-government programs available to help those incapable of making ends meet due to physical or mental disabilities or a sudden financial or medical crisis. In addition, educational and vocational training opportunities abound for serious students willing to put in the effort and elbow grease—those looking for a hand up, not a handout. Potential employers in the real world welcome those who reject the victimhood mentality and accept individual responsibility to work diligently to resolve their problems and contribute to society.

Here are other examples of individuals who experienced life-altering events yet remain grateful and continue to exude joy. They have all been great blessings to myself, my family, and others who know them.

One, whom we have known the longest, is our dear friend, Ruth Perrin. Ruth lived in Chicago with her husband and two young sons, until her husband developed cancer and died. Ruth then moved to our suburb, and her children went to the same school as ours; her oldest eventually played on the same ice hockey team as Jae. Ruth, the daughter of a career Air Force Doctor, became a highly respected researcher with the Department of Veterans Affairs, dealing with statistics involving Veterans' healthcare projects, proposals, and evaluations. Over the years, we kept in touch as she remained in our immediate neighborhood.

Then KJ entered her master's program, and her thesis centered on a research project that involved considerable statistical analysis. Because of her research expertise, Ruth became KJ's technical advisor regarding the research design and statistical evaluation. Psychology research differs from other academic areas, with specific requirements according to the American Psychological Association (APA). Ruth knew the precise format and procedures to follow. When KJ was conducting her research, there was a particular statistical research software program called SPSS. In 2004, the program was so expensive that it was only available in the university statistical lab or to large corporations or government offices. KJ and I used to go to her university lab to run her data. We were always the oldest in the lab, and occasionally, we disagreed on how to utilize the software. The students would enjoy overhearing our "dialogue." But when we later discovered Ruth had a copy of SPSS on her home laptop, she invited us to her home and guided KJ on the correct approach. Ruth's presence in our lives at this time was another "God thing" that enabled KJ to successfully complete her master's thesis with high marks and deep satisfaction. Throughout our adult married lives, this single Mom has always been gracious, hospitable, and inspiring to us both.

Two other individuals who model gratefulness regardless of adverse circumstances are members of our church. One is Steve Lake, a retired school superintendent in our area who was born in Chicago, raised in our suburb of Glenview, Illinois, and has remained there his entire life. His first wife died suddenly from a brain aneurysm at age 51, leaving Steve with a son and daughter in their 20s. Although his world was literally turned upside down in a flash, he remained steadfast in his strong faith and grateful for the support of his extended family and faith community.

His children went to high school with Sumi and Jae, and we have become very close friends with Steve and his second wife, Connie, whom he met at our church years later. Steve is another example of someone who is living his calling. An educator his entire professional adult life and a strong man of faith, he has utilized his spiritual gift of teaching to conduct men's faith-based conferences. He has published a book on spiritual gifting. Like my friend Bud Domagata, he has inspired me to write a book, as I have seen them put pen to paper on their own books. Throughout KJ's and my time knowing Steve and Connie, we cannot think of any two other people we would rather be with for their joyful nature and servant hearts.

The other church member is Jacob Lee, who, with his wife Susan, has an overwhelming aura of constantly serving others 24/7, regardless of their circumstances. Jacob is an immigrant from Hong Kong, a father of two children, and a pharmacist acting as a consultant for firms in the pharmaceutical industry. Jacob was involved in a lawsuit for many years that consumed time and considerable resources unsatisfactorily. Yet, one would never know he was going through such an experience because of his and his wife's always cheerful and serving nature. He is a member of my men's small group and always asks me at our regular meetings, "General, how are you doing today?" knowing full well that he will hear my mantra in return.

Jacob and Susan Lee are among the finest people we know, as they remain grateful for their faith and for living in a land with unique freedom and opportunity. They are always the first to volunteer at church in numerous capacities, helping those in need. Although not a Veteran, Jacob has been an active volunteer with Military Outreach USA, providing much-needed internet and social media skills. Because we are many years older than them, when the COVID-19 epidemic began, Jacob and Susan immediately volunteered to buy groceries and other essentials for us and delivered food on occasion. Jacob has also purchased items for us at the hardware store and offered to shovel our driveway. They do all of this with gratitude and the resultant joy.

I would be remiss if I did not acknowledge the influence of an individual who has been instrumental in the progress of Military Outreach USA and in guiding literally thousands in their life journey, including yours truly. He is Tim Hetzner, former President and CEO of Lutheran Church Charities (LCC). Though their headquarters is in Northbrook, Illinois, they have a nationwide network of houses of worship and organizations that demonstrate their Christian faith in action by serving people and communities in crisis. Through Military Outreach USA, I have been blessed to work with all denominations. I initially met Tim through the active participation of Lutheran Church Missouri Synod (LCMS) in helping our Veterans. His K-9 Comfort Dog Ministries employ beautiful, highly trained golden retriever dogs to provide comfort and support to those suffering from tragedies caused by humans and nature.

For example, they are invited to bring their dogs to struggling families and communities recovering from shootings, fires, floods, and tornadoes. Their network is so extensive that they are typically on the scene within 24 hours, if not less. In addition, they provide food, water, and other supplies as requested. Additionally, they may serve a devastated community by clearing away debris with bulldozers and chainsaws, which they call L.E.R.T. (Lutheran Early Response Teams). All of their work is provided free of charge to the recipients.

The LCC and Military Outreach USA have worked closely to support our Veterans and First Responders in need of comfort as they deal with the invisible wounds of their service to our nation and communities. If that were not enough, Tim also produces a daily devotional that is both instructive and inspirational. Speaking personally, Tim has been a mentor and strong counsel regarding how Military Outreach USA can better serve our God and country. LCC has also provided us with an office in the past and permits us to use their address and office and postage machines. Tim's generous support and counsel made us far more effective as a nonprofit.

I must mention one final strong man of faith who personifies gratitude and patriotism. I have never met him in person and only became aware of him in the past four years. But in light of our shared values and similar age cohort, I consider him a brother in many ways. I have spent hundreds of hours with him via his "Fireside Chats" and through his organization's five-minute videos. You may have surmised that I am referring to Dennis Prager and his Prager University, or PragerU, internet programs. Regardless of your political persuasion, Mr. Prager, or Dennis as he prefers to be called, presents logical, fact-based discussions on various subjects — religion, economics, history, education, health, government, politics, science, and others. These presentations involve world-famous recognized authorities in their respective fields.

Prager University's mission is to educate the public, especially the youth, on matters critical to both individuals and our nation. I stumbled accidentally on one of Dennis' five-minute videos but was so absorbed by their lessons that I continued watching more for the next two hours. Since PragerU is a nonprofit, their videos exist solely on contribution yet have received over 5 billion views worldwide. Their productions have expanded into many other areas, including book clubs for adults and children, and education curricula for K-5, with more to come. I invite you to go to PragerU.com and check it out for yourself. There is a saying that you can't teach an old dog new tricks, but this old dog has learned immensely from PragerU. Being a lifelong student is both a benefit and a responsibility that comes with age. How you apply your learnings to your experience is what produces wisdom.

Regardless of my admiration for PragerU's educational contributions, I also respect Dennis as an individual. Dennis is an Orthodox Jew, and the Bible is his foundation. He is well-traveled and experienced, having been to over 130 countries worldwide, and counting. But most importantly, he is grateful every day to be an

American.

If this last discussion regarding PragerU sounds like a commercial, it is. But it is unpaid and unsolicited. I offer it to you as both an opportunity and a challenge — to be open to respectful discourse and to share if you find my comments accurate.

What do those who appear in this chapter have in common? They are the type of people you want to spend time with, in good and bad times. You always leave their presence feeling better and looking forward to being with them again. They are people grounded in faith and strengthened by their family, friends, and community. They are thankful to be living in the land of freedom and opportunity. They epitomize what it means to be a grateful American.

I conclude by expressing my appreciation to you, the reader, for taking the time and effort to read this book. Profits from its sale will benefit our church and Military Outreach USA. May God bless you.

EPILOGUE: OVERTIME
MAKING A DIFFERENCE?

In the late summer of 2021, I was giving the invocation for the Allen J. Lynch Foundation fundraising event. Prior to the event, my wife and I were in the hallway talking to Al, Andrew Tangen, who is a Navy combat Veteran and Superintendent of the Lake County Veterans Assistance Commission (one of the best in the country), and Karyn Davidman, the executive assistant for Congressman Brad Schneider of the Illinois 10th District and my Congressman, and a strong supporter of our Veterans. I mentioned my work with the Japanese-American World War II Veterans, and Andrew piped up and said, "General, you need to write a book."

Al immediately added, "Yes, General, you should!" Karyn completed the refrain and also voiced her agreement. I responded something to the effect that I didn't think anyone would be interested. There was instant disagreement. Andrew volunteered to help me write it, and Al, who has written his own book entitled *From Zero to Hero*, said he would help get it published. I asked if they were serious about this. They both affirmed and said yes, so I said I would consider it.

The next day, I emailed Andrew and Al to confirm if they were really willing to help publish my book. They both immediately responded yes. If I had their support, this book would be a worthwhile pursuit. I then replied that I was in, and my book-writing journey began.

If you haven't noticed, I am a very logical guy, and approach life's challenges in a methodical manner. I study the situation, devise a plan, and then execute that plan. Accordingly, I followed the same methods from the moment I decided to write this book. First, I asked myself, who was my target audience? Four main groups became the clear answer. Firstly, people of faith, or those seeking it, can connect with my story of faith. Second, the military community would enjoy hearing about my Army career. Third, I hope that minorities (not restricted to skin color) will be inspired by my legacy as the first Asian American to command an Army Division. And finally, I hope that the youth can learn from my story and values. Many readers will fall into more than one of these categories and, together with others who do not, will hopefully find it educational, encouraging, entertaining, or inspirational. The book has done all these things for me in my writing process, and I pray that it has made a positive difference in any reader's life.

During the one-day healing retreat we conduct for Veterans and family members at the Bellarmine Jesuit Retreat Center, we review a daily prayer exercise called the *Examen*, created by St. Ignatius Loyola. He was a respected commander in the Spanish Army before he had a spiritual revelation while recuperating from serious wounds. I was not raised Catholic, so I did not know anything about the Examen. I have incorporated a portion of the Examen in my daily prayers, so I review the previous 24 hours and see where God appeared and how my behavior reflected God to others. As a result, I wake up every morning excited to see what God will do in my life. Sure enough, every day, I receive some kind of daily encouragement. Your time and effort to read this book are one of these encouragements.

As this book was awaiting publication, I was honored with a special Commendation from my ancestral country of origin, the government of Japan. They invited me to a dinner at the official residence of the Consul-General of Japan to recognize my half a century of contributions to promote positive relationships between the nations of Japan and the United States of America — militarily, diplomatically, and culturally. At the ceremony, I declared and accepted "this honor on behalf of the previous first generations of Americans of Japanese Ancestry, or AJA's, the Issei and Nisei, who, by their sacrifices and personal example, established the foundation upon which all future generations of AJA's have benefitted. Their reputation and contributions to our nation opened up the doors of equal opportunity to those such as myself and others." My full remarks are attached as Appendix K[13]. I'm sure my father and uncle smiled approvingly when I received the honor.

Another "God thing" happened after the completion of this manuscript. For the first 19 years that I voluntarily instructed for the Military Ministry of CRU at the Great Lakes Naval Training Center and taught at the Recruit Training Chapel, our total class attendance averaged between 20 to 50 recruits. Our record attendees were 80 students. With the advent of COVID, our classes ceased for about a year. After resuming our classes in 2022 and moving our meetings to Friday evenings, we saw a revival of interest. As we approached the 20th Anniversary of my teaching and my final class presentation on September 30, 2022, the attendance had grown exponentially, requiring our class to move from small classrooms to the main

---

13 Consul General Award Dinner Remarks — Appendix K

chapel. My last class set an all-time record with an attendance of 1,200 recruits. There was not one empty seat in the chapel; recruits sat on the floor and stood across the entire back of the balcony section. The decades of efforts of our loyal volunteers had borne the fruit of our prayers. As I gave my remarks and gazed over the ocean of faces before me, I thought, "Only God!"

Addressing recruits at the Great Lakes Naval Recruit Training Command Chapel.
Great Lakes, IL. 2022

As my wife and I assist our aging friends and Veterans, we are realistically preparing to move to smaller living quarters, most likely into a Planned Living Community with the total continuum of health care services. Accordingly, while sorting through our possessions, my wife found a priceless treasure — two audio cassette interviews of my mother in the last year of her life. A University of San Francisco researcher interviewed her while studying Nisei in America. During that interview, I learned that my family on my mother's side were samurai. Both of my mother's parents came from samurai families.

My grandfather's elder brother had fought under Saigo Takamori in the Satsuma rebellion against the Emperor. This revelation has brought my samurai journey to a full circle.

To help others learn from my foundational values, experiences, and wisdom granted in my life, I have created a summary document containing short lists of the most influential people, books, speeches, and documents in my life. It is attached as Appendix L[14].

Finally, I am passionate that we Americans have every reason to be proud of the exceptionalism and accomplishments of the USA. Recent polls have shown a significant decrease in community, patriotism, and religion in the United States of America. Having reviewed the strengths and weaknesses of our nation via my personal experiences in this book, I summarize my life-guiding principles that may reverse the current decline in the strength of our country, psychologically, physically, and spiritually. The thoughts are entitled, *The Mukoyama Life Manifesto*, and it is attached as Appendix M[15].

High school students are one of my favorite audiences with whom I enjoy sharing my experiences and values. On November 11, 2022, I had the honor of visiting the John Hersey High School of Arlington Heights, Illinois, participating in their Veterans Day activities. It started in the morning with a breakfast with the students and Veterans from the community. It was immediately followed by a school assembly with an audience comprised of the Veterans and the junior and senior classes. The program included a concert by their state champion concert band followed by the keynote speaker (yours truly), and recognition of special Veterans and students for their demonstrated service and patriotism. The band director

---

14 Most Influential List. Appendix L

15 *The Mukoyama Life Manifesto* is found in Appendix M

invited me to participate, and I introduced our Armed Forces Service Songs; read a poem about our National Hymn, *God of Our Fathers*, as the band played a symphonic poem with the hymn as the main theme; and conducted the band in the final piece, *The Stars and Stripes Forever*. My performing with the band also brought my musical career journey to a full circle from when I played clarinet and saxophone in high school.

On that special day, I met a younger Veteran of our Middle East campaigns who had been deployed six times. Like me, he was an Infantry soldier, and we were both graduates of what we affectionately referred to as "Benning's School for Boys" (Fort Benning, Georgia, the home of the United States Army Infantry). Immediately following the school assembly, we participated in the Hersey School Podcast. Though we had only met for the first time that day, the other Veteran mentioned he had known about me prior and was anxious to meet me. He had listened to the entire Jocko Willink Podcast No. 124 years earlier. Being of Filipino descent, he said it was so encouraging for him to see a fellow Asian American able to succeed in life and the military, and he was determined to follow in my footsteps. When he learned about my family life and community service, he started his own non-profit serving Veterans. Not only was I moved by his admiration, but I was extremely humbled when he said he was honored to meet me. The feeling was mutual. As you can see, the leadership principles of Example, Caring, and Balance pay off. Another life circle was completed.

That weekend was capped off with an unexpected event. Unbeknownst to me, I had been nominated for an award from the Mental Health Association of Greater Chicago, along with numerous notable individuals, honoring our efforts in serving our Veterans experiencing tough life experiences. My American Legion

Post had nominated me to recognize the accomplishments of Military Outreach USA in educating the military community and the public on the invisible wound of war, Moral Injury. My wife and I attended the organization's annual fundraiser, and I received the 2022 Vetted2Live Award. The Association also recognized Military Outreach USA for our book, *They Don't Receive Purple Hearts,* and our decades of work and presentations at universities, medical centers, and the American Psychological Association.

It is a true blessing to experience life's season of reaping. But it is equally rewarding to extend life's season of sowing as far into the future as one is given.

In 1982, my wife KJ was studying calligraphy, and as a gift, she gave me the following poem that she painstakingly penned in a calligraphy font. It is framed and hanging above the sink in our master bathroom, and I read it every day. It goes like this:

"Today is mine. It is unique.

Nobody in the world has one exactly like it.

It holds the sum of all my past experiences and all of my future potential.

I can fill it with joyous moments or ruin it with fruitless worry.

If painful recollections of the past come into my mind or frightening thoughts of the future, I will put them away.

They cannot spoil today for me."

— Anonymous

When I recite it, I have changed the beginning to "Today is Yours." — meaning it belongs to God. I pray that the Good Lord at the end of the 4th quarter will bless me with some overtime. It doesn't hurt to ask — who knows?

A few years ago, I addressed a high school class, and days later, I received a thank you note from one of the students expressing appreciation for my sharing my experiences with the students. At the end of the note, he said that my mantra would be his. He has his faith, his family, and lives in the finest country in the world. He gets it.

APPENDICES

APPENDIX A
The Greatest Love Letter Ever Written
January 3, 2021

A letter does not normally qualify for an anthology, but the world's all-time best-selling, most well-read, and most historically influential writing of all time was indeed a love letter. The author is the Creator of the universe, God, and the recipient is anyone who has read it, in whole or in part, in the past, present, or future. It is the Holy Bible.

It begins, continues throughout, and ends with the theme of love. The universe, the world, and humanity were created as acts of love. The original splendor, wonder, and beauty reflect the loving nature of the Creator. Humanity was created in the image of God to live in perfect relationship with Him in a paradise. His special love for humans was demonstrated in giving humanity free will and a uniquely discerning brain with language ability instead of making fully compliant serving robots.

But sin was introduced into the story by man and woman disobeying God. The ensuing spoiling of nature and society is described in the subsequent stories of both good and bad choices, with God always eventually showing His unconditional love to humanity.

The spectacular gift of love was for God to assume the nature of man, leave His heavenly kingdom, and come down to earth ending in His voluntary sacrificial death via crucifixion to re-establish the broken relationship between humanity and the Creator. And this undeserved grace and forgiveness comes without any personal payment on the part of the reader. Neither does one have to work for or earn this love. It is a gift offering available simply through faith by accepting the invitation of letting Jesus enter one's heart as personal Savior.

His resurrection and subsequent time spent with His disciples and other numerous witnesses confirmed He was indeed the Messiah and Son of God and forever transformed the lives and eternal destination for all who accept Him.

Instead of an ending, the final portion of the letter demonstrates the beginning of the ultimate peace and joy that will result when the Lord returns.

The Bible, as any love letter, needs to be cherished and enjoyed over and over, for the more you read it, the more you learn and enjoy. And most importantly, you know that the writer loves you because God is love.

APPENDIX B
Mukoyama Family Code
August 7, 1985

To our dearest daughter, Sumi: Love, Papa & Mama

General Tenets

- Have faith in and be grateful to God
- Obey the laws of God and of our country
- Always tell the truth (even if you expect to be punished)
- If you do something bad, admit it and apologize
- Be sensitive to other people's feelings
- Respect and obey your elders and teachers
- Be polite and use good manners
- Try your best in everything you do
- Have confidence in your abilities
- Be proud of your Korean/Japanese/American heritage
- Be loyal to your family, friends, and country
- Help, encourage, and share with others
- Be personally neat and clean
- Save your money and spend it wisely
- Appreciate living in the United States of America
- Do the right thing even if it is unpopular
- You can talk with Mana and/or Papa anytime about any question, subject, or problem. (Your ideas and feelings are important to us.)

- Be on time for all appointments
- Write thank you notes for all gifts

Daily Conduct

- Make bed
- At mealtimes
  - Set the table
  - Say grace
  - Sit up straight
  - Don't talk with mouthful of food
  - Eat all that you take
  - Request permission to leave the table
  - Take dishes and cups to the sink
  - Help clean up
- Brush teeth after every meal and snack
- Finish school homework
- Practice musical instrument
- Finish Korean school homework
- Always tell your elders where you are going
- No snacks without permission; drinks only in the kitchen
- Clean up your room
- Help clean and keep clean our house, both indoors and outdoors
- Say prayers
- Go to sleep on time

Prohibited Actions

- Do not brag about your accomplishments or possessions

- Do not be discouraged by failure or disappointments. (Often such experiences in life give you lessons which will help you do better in the future)
- Do not talk back to or interrupt elders or teachers
- Do not abuse your body (by smoking, drugs, or alcohol)
- Do not be jealous of others (rather be happy for them)
- Do not be greedy
- Do not be nasty or try to hurt other people
- Do not disturb other people's property
- Do not waste energy or natural resources
- Do not be a complainer or negative person
- Do not be a quitter (finish what you start)
- Do not cheat in school or when playing games
- Do not talk to strangers
- Do not accept gifts, candy, or rides from strangers
- Do not use bad language
- Do not let anyone touch or see your body's private parts
- Do not play in the street or with matches
- Do not jump on furniture

Additions:

- Write thank you notes promptly for all gifts and favors from others
- Keep good posture; do not slouch
- Always remember that Mama and Papa love you even when we get mad when you break these rules
- Do not go to sleep or leave a family member mad at them
- Always be prompt; do not make people wait

APPENDIX C
What is an Angel?

I picked up the newspaper this morning and read an inspiring article about an elderly Certified Nurse's Assistant in a nursing home who gave special loving care to an Alzheimer's patient two days before he died, and the joy that gentlemen received and the gratitude he demonstrated as a result of her sensitivity to his needs and humanity.

As I was reading the story, my mind immediately envisioned my lovely bride of nearly thirty-five years. Like the woman in the article, she works in an elderly healthcare facility. Because she also is older than most of her peers, she is more empathetic to the residents' circumstances. Additionally, she has a master's degree in Gerontology (the study of aging) and, therefore, has a thorough knowledge of the aging process and the diseases most common among the elderly. But her most important quality is the compassion, love, and respect that she exhibits to her residents. As a result, they adore her and look forward to seeing her every day.

I cut out the article and gave it to my wife saying, "Here is an article that is exactly about your compassion for your residents." What I didn't tell her was how the article again reminded me of how blessed I am to have her in my life.

I thank the Good Lord at the beginning of every day for honoring me with such a wonderful life partner. And it was clearly His hand that brought us together. How else can one explain two people from two completely different cultures, from two different continents (Asia and North America), and two different personalities meeting halfway across the continental United States?

Every day I marvel at her sensitivity to others and serving attitude in virtually everything she does — from getting up at 5:30 AM every morning to prepare tea and a breakfast snack for me because I leave very early for my job even though she doesn't leave until much later for her work — to planning every lunch (I come home during the mid-day) and dinner meal so that I receive a nutritious and appetizing experience (not easy for a picky eater like me).

Her serving qualities are also clearly evident in her work and when we make our weekly hospice volunteer visits. She has arranged for special recognition for residents at her facility whose life stories would otherwise have gone unnoticed. Our hospice patients always respond to her presence and touch with warm looks of affection, contentment, and appreciation. These examples are how she serves today, not to mention her life-long dedication as a mother, wife, and daughter-in-law, as well as her compassion for those less blessed in life than ourselves.

Finally, there is a special aura of beauty, both outer and inner, which exudes from her whether it be simply walking outdoors or entering a room. Her smile has been my life's sunshine, and the touch of her hand makes me feel ten feet tall. My wife once asked me, "How does one live a life as a Christian?" I told her the answer was easy, that one needs to demonstrate God's love in everything we do. I should have merely told her to look in a mirror. I don't have to go to heaven to see an angel — I see one every day.

APPENDIX D
Mukoyama Missives

D ate: 15 September, 1989
Subject: Open Communications,

The purpose of these periodic letters is to convey information of critical interest to the women and men of the total 70th Division team — reservists, civilians, AGR's and AC members. I will expect all members of this command to be knowledgeable about the contents of these letters.

It has always been my policy as a commander to maximize communications up and down the chain-of-command. It is virtually impossible for me to lead effectively without feedback from those who must live with and carry out my directives. I have had the good fortune throughout my career to serve with exceptional soldiers who have made recommendations to improve or just plain correct some of my ideas.

Make no mistake. I am confident in my decisions because I develop policies after conferring with my subordinate field commanders, staff, and noncommissioned officers. However, all good intentions and detailed coordination aside, a division policy occasionally might not be applicable to a certain individual or unit due to an unusual situation. It is at that moment that I expect the leadership of the 70th to stand up and be the "heat shield" for their troops and request reasonable relief. All such requests will be given a fair hearing.

In this regard, after four months as your commander, I have now visited every major subordinate command of the Division, both at home station and at annual training. And I have personally interviewed dozens of our enlisted soldiers on a one-on-one basis in an effort to assess unit and leader fitness. Those areas brought to my attention for correction have been vigorously pursued.

I am pleased to report to you that our Division is serving our great nation well. I have been privileged to observe our soldiers at Fort Riley, Fort Jackson, and Fort Benning. Without exception, our active component counterparts spoke glowingly of the women and men of the 70th. You have truly maintained our reputation as "TRAILBLAZER SHARP."

I am absolutely convinced that we could execute our mobilization mission today in a totally professional manner and I have conveyed that message personally to MG Spigelmire, Commanding General of Fort Benning.

As I have mentioned to you on previous occasions, I am privileged to command such a fine group of dedicated, professional citizen-soldiers. I would ask for your continued prayers for the welfare of our Division and our nation.

TRAIN TOUGH!
James H. Mukoyama, Jr.
Brigadier General, USAR

To the Members of the 70th Division
Date: 21 January, 1991 (0715 hours)
Subject: On The Trail

The time has come! The moment for which we have trained has arrived. I am in Washington, D.C. today and have just received the order to mobilize elements of the Division. By the time you read this Missive, you will be aware of which units have been mobilized.

The Trailblazers are now officially, as a unit, participants in our nation's Operation Desert Storm. We are honored to be the first Infantry Training Division to be mobilized. This clearly demonstrates the high esteem which the Department of The Army and Fort Benning have in our abilities and professionalism.

The days ahead will be demanding, but I am supremely confident in our premobilization action, our plans, our commanders and staffs, our full-time support members, and our soldiers and their families. For those members mobilizing, I can assure you that our Family Support Programs have my highest priority.

These are critical times for our nation and for our Army. We, the TRAILBLAZERS of the 70th, have been called — and we are ready. I would ask for your prayers to the Good Lord for his continued blessings on our families and our great nation.

TRAIN TOUGH!
James H. Mukoyama, Jr.
Major General, USAR

To the Members of the 70th Division
Date:14 April, 1991
Subject: Blazer Sharp!

The Secretary of the Army has announced my appointment as Deputy Commanding General (U.S. Army Reserves) for the Training and Doctrine Command, which is headquartered at Fort Monroe, Virginia.

As I take leave of the TRAILBLAZERS, I view, with great pride, the finest Training Division in the entire United States Army. Others may point to higher statistics, but no other Training Division stood as many soldiers or as many units to the colors as did the 70th. And your performance was magnificent, as has been attested to by the personal accolades of the Chief of Infantry and the Commanding General, TRADOC.

Next to our splendid accomplishments during mobilization, I take greatest pride and satisfaction in our achievements in retention during the past two years. In each year, you have established new records for retaining our fine soldiers. This is the clearest indicator of caring and ethical leadership.

Our challenge now is to maintain our high state of readiness through strong sustainment training programs and immediate attention to correct the deficiencies uncovered by the mobilization process.

In my previous Missive, I asked you to "...keep faith with our God, our leaders, and our great nation." The Good Lord has answered our prayers and our cause has prevailed. May He continue to bless our families, our division, and our great nation.
I bid you farewell.

TRAIN TOUGH!
James H. Mukoyama, Jr.
Major General, USAR

APPENDIX E
Moral Cowardice Endangers Soldiers
David H Hackworth

Courage is as much a part of soldiering as gunpowder. But having guts isn't just about charging the enemy. It's also about standing tall against wrongdoing and fighting for what's right.

Moral courage is in short supply in today's armed forces. The Navy's Tailhook scandal and the Air Force's overrun scam with its $1.5-billion C-17 transport aircraft are proof that courage to fight misconduct is lacking in the top ranks of these two armed services.

A recent decision indicates the U.S. Army may be infected with the same disease. Army Gen. J.H. Peay testified recently that he approved a sweeping overhaul of the Army's Reserves, putting combat functions in the National Guard and giving the U.S. Army Reserve support missions. Peay admitted this decision was made without a "cost analysis," nor were operational tests conducted — an exercise in olive-drab stupidity, like buying a fleet of new tanks without asking the contractor for the cost or specs, and not bothering to check whether the tanks will be Army green, Navy blue or shocking pink.

Sources allege Peay locked up a squad of Guard and Reserve brass, twisted arms, privately threatened one general for playing footsie with a fraulein, and in the end, cut a politically expedient deal good for the porkers but bad for the grunts in the foxholes.

Peay, defending his "Look Mom, no homework" decision said, "There was a general feeling there were efficiencies to be achieved by going this way."

He gave an unprepared sophomore student's quick answer that "$100 million" would be needed to do the deal, while others who did their accounting say this figure could run to a cool $1 billion. Still, $100 million or $1 billion is a hell of a lot of taxpayers' money not to put a pencil to.

Gen. James Mukoyama, a part-time warrior who drills with the U.S. Army Reserves, is ballistic concerning Peay's poorly reasoned plan. He said, "This reorganization will endanger soldiers' lives, degrade readiness and waste taxpayers' money."

As "Minutemen" have since Paul Revere's famous ride, Mukoyama serves because he's a patriot. The bantamweight two-star general, a much-decorated Vietnam War hero, sounded off at a recent congressional hearing. He called the decision "haphazard" and "nonsensical," and said that "politics should not be allowed to override military considerations."

Mukoyama is right. I've commanded Guard units during war and peace and observed both Reserve components in the field, from the end of World War II to Desert Storm.

The Guard's a political organization filled at the top with "old boy" appointees. Its function is to be a cash cow for its "home" state, not to fight.

The Reserves, a different breed entirely, are held to the same standards as the Regular Army; top promotions are not made because you know somebody in the governor's mansion, but because you're tactically competent.

Future wars will be fought on a come-as-you-are basis. The

whistle will blow and our active and Reserve forces will launch, ready or not. They won't have years to hide behind our oceans while getting ready, as they have since 1917. The Guard's major combat ground units, though gallant in WWII, will have been dismal since. They're as obsolete as the horse and the saber.

The Guard is filled with well-meaning folk who are remarkably ill-prepared for battle, while their top brass know more about attending meetings than war-fighting.

The three Guard brigades called up for the Gulf theater were never deployed because they couldn't get their act together. A Regular Army major who trained once said, "They were worse than the Iraqi Army."

On the other hand, the Reserve units put in a stellar performance. Peay should ask his Regular Army Reserve / Guard advisers, past and present what they think, and prepare for an earful.

Congress should order an accounting of what Peay's changes will cost and pit Reserve and Guard combat units in mock battle to see who's best. Along the way, Mukoyama should receive a medal for displaying rare "above and beyond" moral courage for telling the truth.

It might start a trend to stop the corruption at the top that's tarnishing our Generals' and Admirals' brass.

APPENDIX F
Understanding
Sumi Mukoyama

Three years ago, my father sat our whole family down at the kitchen table with an announcement. "I am going to testify against the Active Army in front of Congress," he started off the conversation, "and your Mom and I wanted to let you know what may happen to my military career." Dad explained the entire situation to my brother and me.

Basically, the Active Army decided to close a good number of Army Reserve bases around the nation; the National Guard would expand and "take the place" of the laid-off Reservists. The reason for this downsizing was the fact that the Soviet Union had collapsed, and therefore no longer posed a threat to the United States as a superpower.

When the Active Army announced their decision during a press conference, they claimed that the Reserve fully backed the Active Army's decision. Due to my father's rank and station at the time, he knew that this statement was false and that the Active Army was making a mistake by replacing the Army Reserves with the National Guard because in his opinion, the nation would not be properly prepared for an international crisis. "Since I am up for a new station next year, the Active Army may not grant me my application because of my testimony," he finished.

My brother asked, "Does that mean you're going to have to retire next year if the Army doesn't give you the station?"

Dad paused a moment and said, "That's exactly what it means, son."

Jae and I looked at each other for a moment. All of our lives, our father had been in the Army. When we were little kids, Dad would always be gone on the weekends because he had to fly every Friday to his station, whether it was in Michigan or Virginia.

We were used to him not being around; in fact, we took it for granted. Nevertheless, we knew our father well enough to understand that he was a moral man — that he always tried to do what he felt was right. And so, we told Dad that we understood the situation and that we fully backed this decision to jeopardize his military career and testify before Congress. With that, my father's prediction came true. The next month he testified, and the next year he was forced to retire.

It was only during the middle of his retirement dinner that I truly understood what a bold move my father had made. He had been in the Army ever since college. He had voluntarily served in the Vietnam War and worked at the demilitarized zone in Korea. My dad had dedicated over twenty-five years of his life to the Army, and now that same Army was silently forcing him into retirement. I had never realized how much the Army meant to him, and I did not understand the implications of his testimony in Congress the year before. Sitting at that round table and listening to my father give his retirement speech, I held a deeper respect for Dad than I had ever felt before. I was proud of him.

Even though my dad probably understood quite well what his testimony would do, I believe that deep down inside he was hoping for a storybook ending. Hidden within his heart, he wanted the Active

Army to right its wrong and to continue his military career as a Commander at a new station. Underneath his realistic notions, his idealistic beliefs hid with quiet expectations. Nevertheless, my father knew the ways of the Army and he knew that his testimony would end his career. The situation was unfortunate; in fact, acrimonious. Yet he understood the consequences of every action that he had made.

I have realized the political issue — the Active Army's decision to cut back on the Reserves — though important nationally, is not nearly as crucial as the personal ethical issue — standing up for what's right. For me, this understanding only occurred at my father's retirement dinner. When my dad gathered us around the table and told us of his intentions, I thought that I understood what he was saying. Truthfully, I did not expect the Army to force him into retirement: he had a virtually spotless record, was in good physical condition and, as a major general, had proven his skill as a leader of men. It was not until his retirement dinner that I appreciated his bravery, and that his testimony before Congress may have been his most courageous act as a military man.

I also realized just how much of his life he was letting go. Even though I may have understood deep down inside what I was saying to my father before he testified, this understanding did not reach the surface of my reasoning. I knew, yet I didn't. I understood everything, yet I understood nothing at the same time. This way of understanding is how life will be, not just for me but for everyone. Inevitably, the decisions that one makes in one's life always comes down to an ethical issue. My father chose the path that he thought was right; even though this action cost him his military career. Because of his bravery, I have realized that life continues to expand understanding, shedding light on every question and opening up doors that had been previously closed before. I hope that my journey of understanding never ends, and that I will someday be able to follow my Dad's footsteps.

APPENDIX G
MOTHER'S DAY

Mothers Day 2003

My Dearest KJ,

As I ponder the blessings of our family, I am filled with wonderment for your being the Mother you are. Sumi, Jae, and I are certainly not deserving of your presence in our midst.

Motherhood is love, exhibited by caring and sacrifice, and I begin each day thanking the Good Lord for giving our family the honor of being able to call you "Mama."

You have cared for us starting with your bringing Sumi home from Korea and nursing Jae to robust health after he arrived to complete our family. You were always there for the children, at school, at church, at sporting events, at skating lessons, and at music lessons, recitals, competitions, and concerts. You did all of this while being the primary caregivers for Obaachan and Ojichan and you kept our home beautiful, organized, and spotless without outside help. And you provided chauffeuring and wonderful meals which were always well-planned, healthy, and appetizing. And you were always supportive of all of us in all of our endeavors, especially me as I pursued my career in service of our nation. Never once did you complain that your load was heavy.

Through all of this, you continually sacrificed your personal goals and convenience for all of us. You came to live in the extreme four seasons of Chicago from the sunny climates of Northern California and Arizona. You postponed your university degree pursuit to devote your time and energy to the children, my parents, and me. You were always there for Sumi and Jae when they came home from school, my parents when they were most in need, and for me when I needed your love, advice, encouragement, and support.

You have been the strong foundation for our family and we all know that and love you more than you will ever realize, even though we don't demonstrate that enough to you.

This Mother's Day, I wanted you to have that demonstration in the form of this letter.

I Love You Forever,
Jim

Mother's Day 2020

The Person I Admire (and Love) the Most In
This World — Kyung Ja Mukoyama
By James H. Mukoyama, Jr.
(her blessed husband)

- Inner beauty — life values of honesty, integrity, empathy, compassion, patience, courage, generosity, and learning
- Physical beauty- her smile is my sunshine; touch of her hand makes me feel 10 feet tall; just being with her gives me joy
- Selfless serving- sacrificed her life, educational ambition, schedule for others/ supported Jim's Army career/ provided children's Mother at home experience/ agreed to Jim's parents living in the same home and caring for them for 22 years until they passed away.
- Humility- always giving others credit for her personal successes
- Courage & Perseverance- regardless of life obstacles never gives up
- Accomplishments- immigrant, wife, Mother, Daughter-In-Law, student, employee, friend, mentor, volunteer, patriot, Christian, servant
- Faith- deep and always growing Christian faith
- Grateful attitude- for God's love, grace, and blessings/ living in America/ our marriage/ our children/ health/ education/ career utilizing her Gerontology education/ volunteer opportunities in retirement
- Nurturing love- for all living things: humans, animals and plants
- Empathy- for those suffering: Relationally, psychologically,

physically and financially/ strangers detected this quality and would approach her for help (in restaurants even)

- Healing instinct- perceives suffering in others and takes action, e.g. when I am physically ill
- Encouragement- always lifting those up in times of need and discouragement
- Grace- forgiving and still loving me despite all my faults since 1971

APPENDIX H
"Systemic Racism Is Not Alive in America"

J une 6, 2020

On the recent 76th D-Day Anniversary, I was moved to address the current war for the future of our nation centered on the proposition that the United States of America is systemically racist.

I believe that all reasonable citizens agree that the death of George Floyd was a totally unjustifiable murder. And it is our constitutional right to assemble and protest. It is not our right to riot and destroy private and public property, thereby hijacking the purpose of peaceful protests.

My issue is with the ever-growing promoting by our media and educational institutions that systemic racism and oppression is inherent in American society. Unquestionably segments of our society encounter discrimination daily. Unfortunately, society is comprised of sinful humans, both in and under authority, and we are simultaneously involved in spiritual warfare with the forces of evil.

I would submit that racism is not "systemic" in our nation because the opposite is true from the very inception of our republic. Rather, the systemic nature of our society is freedom founded in

God-given inalienable rights.

I am the son and husband of non-white immigrants who came to America precisely because of its promise of opportunity. I have lived for 75 years and I can witness to opportunity and racial societal improvements in my lifetime. I grew up in the inner city of Chicago. Neither of my parents had a college education and we had a small family retail store business. We always lived in an apartment building. Because of my faith and family, I never felt poor. Every Sunday we would dress up and walk as a family to church. My father emphasized the responsibility of being a grateful, loyal American citizen because of our privileges of freedom, education, and opportunity. He said to be proud of our ethnic heritage, but to be prouder of being an American citizen.

Because of that moral foundation, I have been blessed to be the first in my family to graduate from college and had rewarding military, civilian, and community serving careers.

Further reflecting on race relations, when I was a kid, the odds of me becoming a two-star general in the Army were virtually nil. I have seen an African American elected President and re-elected. Interracial marriages today are accepted societal norms. My Army career demonstrated how a society functions with equal opportunity. The color of your skin (and gender) was irrelevant – we were all Army green. Promotion was based on ability, work ethic, and results. We received equal pay for equal work. I have seen an Asian American appointed as the Army Chief of Staff and an African American appointed as the Chairman of the Joint Chiefs of Staff.

And America has not had a blind eye to our past, taking numerous actions to improve opportunities for deprived communities. These have included early childhood education, school nutritional programs, scholarships, and passing legislation addressing redlining and job-hiring discrimination. And the United States is the only nation in the world that has had a civil war that cost over one-half million military casualties to abolish slavery.

And the church has acted. The abolitionist movement was led by people of faith. The civil rights movement of the 60s was supported by clergy throughout. Martin Luther King was a pastor.

I have experienced prejudice in my life, often life-changing, but throughout I received strength from my trust in God. I have also experienced people in my life who provided me help, hope, and encouragement, and the vast majority were not people of color, but white. These experiences are, in fact, the systemic nature of our American society.

The false mantra of systemic racism as the core nature of our nation should be rejected. If one is constantly told you are a victim, you naturally feel anger, hopelessness, and resentment. However, if you understand that America offers true opportunity, you will be grateful.

James H. Mukoyama, Jr.
Major General, U.S. Army – Retired

APPENDIX I
One Foot on the Banana Peel

The Power of Giving Thanks
Air Force Chaplain Dr. Jerry Hardwick
November 23, 2017 at 8:34AM

In everything give thanks; for this is the will of God in Christ Jesus for you. (1Th 5:18 NKJV)

There is a saying: If you can't sleep, count your blessings, not sheep. On this great American Holy Day we celebrate our Judeo-Christian tradition. As a nation we take a Sabbath rest, gather together with family and friends and worship God with sacrifices of Thanksgiving.

Obviously, Maggie and I have something very special to thank God for... This time last year we were giving thanks in Taos, New Mexico, as guests of our Christian brother and sister, Tom and Ruth Thomson. We took a snowy hike up to 1,000-plus-foot elevation Lake Wheeler. Thinking it was just the cold air, I started to get a sore throat — the first symptom that the strange lump on my throat I noticed just a month before may be something else.

By the time the doctors had a good diagnosis, it was a bad diagnosis with a terrible prognosis: "Six months to a year if you refuse our protocol." And so, I was placed into hospice.

At this writing, we just returned from Chicago, where I was

honored by CRU Military for pioneering the Great Lakes Naval Training Center fifteen years ago. I was overcome with emotions to hear one-by-one, testimonies of how Maggie and my lives have touched them, and so many hundreds and thousands through these staff and volunteers, many of whom I had recruited back then. Now, some are full-time ministers themselves. Too many to start naming them all, but one I need to single out: Jim and K.J. Mukoyama, MGen. U.S. Army – Retired.

The Mukoyamas hosted us in their lovely home and treated us like royalty! Upon arrival, he presented me with a 3-ring binder, with a picture of an old Banana Peel on a table in front of their Family Bible. In it was every Banana Peel I've sent out. Jim and I inspire one another to love and good works (Heb. 10:24). He says my life inspired him to go into ministry full time. Now he is the founder and CEO of a very important ministry to Veterans called Military Outreach USA (Take a look at this, and maybe be inspired to get involved).

So, this Thanksgiving, give thanks in everything. We can even give thanks to God in this battle we're fighting. Thanking God in the battle, not for it. Why? Because it really, most sincerely is true that God can be glorified in this. He can indeed turn bad things into good. So give thanks in the battle(s) you are in, and watch God be glorified in you!

And we know that all things work together for good to those who love God, to those who are the called according to His purpose. (Rom 8:28 NKJV)

<December 8-18 it's back to Honduras to do "Los Ciolos Proclaman" the Spanish version of my entire Biblical Astronomy

presentation called *The Heavens Declare*>

Remember! You too have one foot on a banana peel! *This is the day that the Lord has made! Will you rejoice and be glad in it? (You have to exercise your will to rejoice.) For the joy of the Lord is your strength! Be strong in the Lord and the power of His might! Live today for all it's worth! Better is one day in his presence than thousands elsewhere! (Ps. 84:10)*

In His Majesty's Royal Service, (1Pet.2:0; Rev. 1:6)

APPENDIX J
United States Military Challenge
Coin Collection

1. Commander-in-Chief
2. Department of Defense
3. Deputy Secretary of Defense
4. Assistant Secretary of Defense
5. Joint Chiefs of Staff
6. Department of the Army
7. Army Chief of Staff
8. Sergeant Major of the Army
9. U.S. Army Reserve
10. Chief, Army Reserve (LTG)
11. Chief, Army Reserve (MG)
12. U.S. Army Reserve Command
13. Training and Doctrine Command
14. Forces Command
15. Army Materiel Command
16. U.S. Special Operations Command
17. Supreme Allied Commander Europe
18. Central Command
19. Southern Command
20. U.S. Army South
21. U.S. Army Europe (7th U.S. Army)
22. U.S. Forces, Korea

23. U.S. Pacific Command

24. U.S. Joint Force Command

25. Stabilization Force, Sarajevo

26. I Corps

27. III Corps

28. V Corps

29. VII Corps/Desert Shield

30. IX Corps

31. XVIII Airborne Corps

32. Multi-National Corps Iraq

33. U.S. Army Pacific

34. 1st U.S. Army

35. 2nd U.S. Army

36. 3rd U.S. Army

37. 4th U.S. Army

38. 5th U.S. Army

39. 6th U.S. Army

40. 8th U.S. Army

41. Combined Arms Center

42. U.S. Army Special Forces Command (ABN)

43. U.S. Army Civil Affairs and Psychological Operations Command (ABN)

44. U.S. Army Accessions Command

45. Civilian Aide to the Secretary of the Army

46. Army Surgeon General (All Medical Related Branches)

47. Army Training Support Center

48. Troop Support Command

49. Soldier Support Center (Adjutant General and Finance

50. Aquisition Branch

APPENDIX K
Consul General Award Dinner Remarks
6/10/22

Consul-General Tajima, Honored Guests and Family Members:

Thank you to all for coming. You are all very special people in our lives and K.J. and I are very grateful for your presence. It is a distinct honor and privilege to be recognized with this award ceremony and dinner presented by the consul-general of Japan.

I accept this honor on behalf of the previous first generations of Americans of Japanese Ancestry, or AJA's, the Issei and Nisei, who, by their sacrifices and personal example established the foundation upon which all future generations of AJA's have benefitted. Their reputation and contributions to our nation opened up the doors of equal opportunity to those such as myself and others to follow.

I would like to recognize members of this generation present tonight, who are Mrs. Reiko Harada and Mr. and Mrs. Masaru Funai. The values inherent in the Japanese culture of honor, integrity, perseverance, respect for elders, were inculcated in my brother John and I by our parents, Hidefumi and Miye Mukoyama. Settling in Chicago prior to World War II, they were leaders in the Japanese-American community, helping establish the Japanese Mutual Aid Society and the Re-settlers Organization assisting those relocating to the Midwest from the concentration camps of the War Relocation Authority.

And the distinguished wartime record of the 100th Battalion, 442nd Regimental Combat Team and the Military Intelligence Service paved the way for me to become a Major General in the United States Army.

I share any modest success in life equally with my dear wife, K.J. She has stood by me for over fifty years and both encouraged and inspired me with her personal example of sacrifice for others as a wife, mother, and daughter-in-law. While I was often away with my Army duties during my 32-year military career, she took care of my parents who lived with us in our home for the final 22 years of their lives. Her love and wise counsel have kept me grounded and made sure that my head did not exceed my hat size.

The good Lord has truly blessed me. Every day is a great day! I have my faith, my family, and live in the finest country in the world.

Thank you.

APPENDIX L
"Most Influential List"
(in order of priority)

Peﾃople

- Jesus
- My Parents
- My Wife
- Our Children
- Men's Small Group

Books

- The Holy Bible – inspired by God
- Purpose Driven Life – Rick Warren
- Evidence That Demands a Verdict – Josh McDowell
- The Five Love Languages – Gary Chapman
- The Servant – James C. Hunter
- The Gift of Peace – Cardinal Joseph Bernadine
- Heaven – Randy Alcorn
- Live Like You Mean It – T.J. Addington
- RARE Leadership – Marcus Warner and Jim Wilder

Speeches

- Polonius's Advice to Laertes (William Shakespeare, Hamlet Act I, Scene III)

- *Duty, Honor, Country* by General of the Army Douglas MacArthur
- King Henry's St. Crispin's Day Speech (William Shakespeare, Henry V, Act IV, Scene III)

Documents
- The Manhattan Declaration
- The Declaration of Independence
- The Constitution of the United States

APPENDIX M
THE MUKOYAMA LIFE MANIFESTO

The following analysis and proposed actions are urgently needed to reverse the current decline in the strength of our nation — psychologically, physically, and spiritually.

The first breakdown is in the basic family unit. Motherhood is no longer revered and celebrated as a noble and honorable pursuit. It used to be that pregnant women were given preferable treatment; doors were opened and seats vacated in public transportation for these women. Fatherhood is no longer embraced as a worthwhile example of responsibility to the young. The current absent-father generation reflects the negative result of the high incarceration rate of youth, especially boys, without a positive male role model in the home. Community activities generating family involvement are rapidly evaporating, such as scouting programs. This loss has been further exacerbated by the refusal to accept the science of basic biology that recognizes two sexes and the innate differences between the two, resulting in simultaneous emasculation of males and defeminization of females in our society — in other words, an unnatural blending of the sexes.

The second essential degradation in our society is in the pride and gratefulness for being an American, and for the accomplishments of our previous generations and sacrificial contributions in advances for citizens and the prevention of totalitarianism, both in our society and in foreign lands. Our nation eliminated slavery within our country at the cost of 300,000 soldiers. We have lost over 700,000 more service members fighting to prevent the expansion of murderous dictatorships worldwide. When there is a disaster, either caused by nature or by man, the United States has consistently been the most generous in response with humanitarian and financial aid. We Americans have every reason to be proud of the exceptionalism and accomplishments of the USA.

The loss of the first two aforementioned societal values is connected to the most important fundamental societal value, and that is a sense of purpose greater than oneself that comes from spirituality, whatever form that might take. My life experience anchored in my Christian faith resulted in three critical elements for maintaining satisfaction in life and a strong nation: gratefulness, forgiveness, and redemption. Having experienced the extremes of life in peace and war, home and abroad, I have witnessed the best and worst of human nature as well as the serendipity of life. Everyone will experience suffering. The keys to surviving are a belief system that recognizes and is thankful to a supreme creator, instead of relying on our limited personal human abilities (pride); to forgive those who have done you undeserved harm, which precludes the festering harm of spite and revenge; and to believe that there is hope for ourselves to be forgiven and transformed into the special beings that we were created to be.

Less than one generation ago when I was a youth, America was strong and flourishing. As a minority myself, I can attest to the tremendous progress that was in fact accomplished and was continuing to improve. An abandonment of the core values established by our nation's founding fathers — based on what Dennis Prager refers to as "the American Trinity": 1) in God We Trust; 2) e pluribus unum; and 3) liberty — has resulted in the sad state of affairs today. The good news is that it can be reversed. It will take dedication, sacrifice, and courage. May the good Lord bless our efforts to do so.

HONORS

HONORS AND RECOGNITIONS

- President, Student Council, Avondale Elementary School, 1957

- Valedictorian, Avondale Elementary School, 1958

- President, Pilgrim Fellowship Youth Group, Christ Congregational Church, 1959

- Vice President, Pilgrim Fellowship Chicago Association, 1960

- Junior Year Service Award, Carl Schurz High School, 1960

- Recipient, Chicago Tribune Gold Medal Outstanding University of Illinois ROTC Cadet, 1960

- Chaplain, Key Club, Carl Schurz High School, 1960

- Inductee, National Honor Society, Carl Schurz High School, 1960

- First Chair Clarinet, Carl Schurz High School Band 1961

- Principal Woodwind, Carl Schurz High School Orchestra, 1961

- Secretary and Librarian, Carl Schurz High School Band, 1961

- Valedictorian, Carl Schurz High School, Summer 1961

- President, National Society of Scabbard and Blade, University of Illinois, 1965

- Distinguished Military Graduate, Army, University of Illinois, 1965

- Honor Graduate, Infantry Officers Basic Course, Fall 1966

- Recipient, Japanese Ground Self-Defense Force Parachutist Badge, 1967

- Recipient, Republic of Korea Army Letter of Commendation, 1967

- Recipient, Republic of Korea National Police Letter of Commendation, 1967

- Member, Chicago Stock Exchange, 1975

- Member, Chicago Board Options Exchange, 1978

- Member, New York Stock Exchange, 1980

- Inductee, Hall of Fame, Carl Schurz High School, 1984

- Special Commendation Poem, Prime Minister of Japan 1987

- Inductee, Order of Saint Maurice, National Infantryman's Association, 1990

- Inductee, The Honorable Order of Saint George, United States Armor Association, 1991

- Inductee, Military Order of Saint Barbara, United States Field Artillery Association, 1992

- Founder & President, Army Reserve Association, 1992

- Director, National Japanese American Memorial Foundation, 1993

- President of the Board, Heiwa Terrace Senior Residence, Chicago, 1997

- Chairman, Advisory Committee for Minority Veterans, Department of Veterans Affairs, 2005

- Founder, Chairman & CEO, Military Outreach USA, 2012

- Inductee, Initial Class of the University of Illinois Army ROTC Hall of Fame, 2013

- Director, Pritzker Military Museum & Library, 2015

- Lifetime Service Award, Volunteers of America, 2016

- Bicentennial 200 Honor Award Illinois Department of Veterans Affairs, 2018

- Director, Friends of Fisher House Illinois, 2019

- Chicago Illini of the Year, University of Illinois Alumni Association, 2020

- Chaplain, Chicago Nisei Post 1183, American Legion, 2021

- Special Commendation, Consul-General of Japan, 2022

- Vet2Live Award, Mental Health Association of Greater Chicago, 2022

US MILITARY AWARDS & DECORATIONS

- Distinguished Service Medal

- Silver Star

- Legion of Merit

- Bronze Star Medal with "V" Device and Two Oak Leaf Clusters

- Purple Heart

- Meritorious Service Medal with Two Oak Leaf Cclusters

- Air Medal with Two Oak Leaf Clusters

- Army Commendation Medal with "V" Device and Five Oak Leaf Clusters

- Army Achievement Medal

- Army Reserve Components Achievement Medal with Oak Leaf Cluster

- National Defense Service Medal with Oak Leaf Cluster

- Armed Forces Reserve Medal with Oak Leaf Cluster

- Army Service Ribbon

- Overseas Deployment Training Ribbon

- Combat Infantryman Badge

- Expert Infantryman Badge

- Parachutist Badge

- Aircrewman Badge

- Expert Marksmanship Badge – M16

- Two Overseas Bars

FOREIGN DECORATIONS/BADGES

- Republic of Vietnam Campaign Medal with Device (1960)

- Republic of Vietnam Cross of Gallantry with Silver Star

- Republic of Vietnam Staff Service Medal 2nd Class

- Republic of Vietnam Civic Actions Medal with Palm Unite Citation Badge

- Japanese Parachutist Badge